Animals, Feed, Food and People

An Analysis of the Role of Animals in Food Production

AAAS Selected Symposia Series

Animals, Feed, Food and People

An Analysis of the Role of Animals in Food Production

Edited by R. L. Baldwin

Routledge
Taylor & Francis Group
LONDON AND NEW YORK

First published 1980 by Westview Press

Published 2018 by Routledge
52 Vanderbilt Avenue, New York, NY 10017
2 Park Square, Milton Park, Abingdon, Oxon OX14 4RN

Routledge is an imprint of the Taylor & Francis Group, an informa business

Library of Congress Cataloging in Publication Data
Main entry under title:
Animals, feed, food and people.
(AAAS selected symposia ; 42)
"Based on a symposium which was held at the 1979 AAAS national annual meeting in Houston, Texas, January 3-8. The symposium was sponsored by AAAS Section O (Agriculture)."
1. Livestock--Congresses. 2. Animal food--Congresses. 3. Feeds--Congresses. I. Baldwin, Ransom Leland, 1935- II. American Association for the Advancement of Science. Section on Agriculture. III. Series: American Association for the Advancement of Science. AAAS selected symposium ; 42.
SF5.A67 338.1'9 79-27368

ISBN 13: 978-0-367-02154-2 (hbk)

About the Book

Numerous authors have presented analyses of the world food problem and the appropriate role of animals in food production and have drawn qualitative conclusions. However, projection and planning require quantitative considerations, and this volume addresses that challenge. Experts in animal science, farm management, economics, international agriculture, and nutrition elucidate and debate germane issues with scientific rigor. They examine the efficiency and economics of animal production, feed resource availability, interactions between plant and animal agricultures, international trade, resource allocation, roles of animals in developing countries, and the nutritional values and limitations of animal products.

About the Series

The *AAAS Selected Symposia Series* was begun in 1977 to provide a means for more permanently recording and more widely disseminating some of the valuable material which is discussed at the AAAS Annual National Meetings. The volumes in this *Series* are based on symposia held at the Meetings which address topics of current and continuing significance, both within and among the sciences, and in the areas in which science and technology impact on public policy. The *Series* format is designed to provide for rapid dissemination of information, so the papers are not typeset but are reproduced directly from the camera-copy submitted by the authors, without copy editing. The papers are organized and edited by the symposium arrangers who then become the editors of the various volumes. Most papers published in this *Series* are original contributions which have not been previously published, although in some cases additional papers from other sources have been added by an editor to provide a more comprehensive view of a particular topic. Symposia may be reports of new research or reviews of established work, particularly work of an interdisciplinary nature, since the AAAS Annual Meetings typically embrace the full range of the sciences and their societal implications.

WILLIAM D. CAREY
Executive Officer
American Association for the Advancement of Science

Contents

Figures and Tables

Chapter 1

About the Editor and Authors

R. L. Baldwin *is professor and chairman of the Department of Animal Science at the University of California, Davis. His fields of interest are nutritional energetics, digestion and metabolism in ruminants, and the mathematical analysis of animal systems. He is currently associate editor of the* Journal of Nutrition.

Henry S. Bayley, *professor of nutrition at the University of Guelph, Ontario, Canada, is a specialist in energy metabolism and feed evaluation and has recently been concerned with the energy value of rapeseed products, the use of molasses in swine feed, and the utilization of nutrients for growth and development in young animals.*

R. F. Brokken, *an agricultural economist for the U.S. Department of Agriculture in Oregon, is particularly concerned with the economics of grain and roughage use as beef rations and in cattle finishing.*

A. C. Bywater *is assistant professor of animal science, University of California, Davis. His field of interest is analysis and management of animal production systems, including computer simulations of farm decisions and production and feed usage in California livestock and poultry operations.*

Ludwig M. Eisgruber, *professor and head of the Department of Agricultural and Resource Economics, Oregon State University, is a specialist in farm management and production economics and policy. His recent work includes determination of management options for beef producers and forecasts of beef production.*

Larry Martin *is an associate professor of agricultural economics at the University of Guelph, Ontario, Canada. His area of specialization is agricultural market and policy* *analysis and his recent publications include econometric simulation of the North American livestock industry, agricultural policy analysis and product supply response, and impacts of commodity futures markets.*

Robert E. McDowell, *professor of international animal science at Cornell University, has focused on the constraints to animal production in warm climate regions of the world. He is the author of* The Improvement of Livestock Production in Warm Climates *(W. H. Freeman, 1972) and has written numerous articles on this topic as well. Other research interests include ruminant products other than meat and milk and problems of small farmers in developing countries.*

Karl D. Meilke *is associate professor of agricultural economics at the University of Guelph, Ontario, Canada. His research has centered on agricultural market and price analysis. He has published econometric simulations of the North American feedgrain and wheat industries, agricultural product supply response, and agricultural policy analysis.*

Nathan E. Smith *is associate professor of animal science, University of California, Davis. His fields of interest include dairy cattle nutrition and management, ruminant metabolism, nutritional energetics and applications of systems analysis in animal agriculture. His current research emphasis is in by-product utilization in dairy cattle.*

James K. Whittaker *is assistant professor of agricultural and resource economics, Oregon State University. He is a specialist in production economics and acreage supply.*

Vernon R. Young *is professor of nutritional biochemistry at MIT. He has a broad research background in human nutrition and a special interest in the interactions of nutrition, aging and nutrient metabolism. He is director of the American Board of Nutrition and is a member of the editorial boards of the* Journal of Nutrition *and the* American Journal of Clinical Nutrition.

Preface

The concept of this symposium was originated by Vice President J. B. Kendrick of the University of California, under the title "From Chuck Wagon to Burger King". The concept was accepted by AAAS reviewers. Development of a title more suitable for key word indexing was requested. The title "Animals, Feed, Food and People" when taken intact, hopefully conveys something of the conceptual flow of the symposium. The title may still fail in terms of key word indexing. The key words are broad in their implications but, so also, is the symposium.

The concept of the symposium arises from the fact that although numerous authors have reported analyses of the appropriate role fulfilled by animals in the production of food, only rarely are all issues appropriately identified in such analyses. Most careful analyses support the conclusion that use of feed grains in animal agriculture will decrease in the future, at least in some areas, but that products of animal agriculture will continue to contribute significantly to the world food supply. Unfortunately, this is a qualitative conclusion not overly useful for purposes of projection or planning. Also, as might be expected when available analyses are not comprehensive and quantitative, many conflicting conclusions and statements have been made regarding the contributions of animals to the human food supply. Both under- and over-estimates of the appropriate role of animals in food production could be compromising to our food supply.

The real challenge is quantitative consideration of the many relevant data and issues and development of methods for determining optimal balances and interactions between plant and animal agriculture for the many differing agricultural lands and circumstances in the world. This challenge has not yet been met. The symposium was organized to identify issues often overlooked and review progress toward putting the role of animals in food production into quantitative perspective relative to allocation of the world's land, water, energy and labor resources. Issues emphasized include alternate strategies utilized in food animal production, biological and economic efficiencies of alternate production strategies, identification of total feed resources available and used for animal production, interactions between plant and animal agriculture, implications of international trade, competitions between humans and animals for feed grains, allocation of feed resources among animal species, social and economic aspects of resource allocation to animal production, roles of animals in developing countries and nutritional values of animal products.

An underlying theme in the symposium is the contention that in addressing the world food problem we must utilize all available resources in an optimum fashion for the production of human food. Achievement of such an optimum requires rigorous, comprehensive, quantitative analyses of interactions among physical and biological resources and social, economic and technological forces. Solutions proposed on the basis of limited data and without consideration of the complex interactions among plant and livestock agriculture could compromise our food supply.

R. L. Baldwin

Animals, Feed, Food and People

An Analysis of the Role of Animals in Food Production

A. C. Bywater, R. L. Baldwin

1. Alternative Strategies in Food-Animal Production

Introduction

Perhaps one of the most fundamental and complex of modern issues has been perennial concern over the many elements of the food versus population equation. The problems of world food shortages have been all too obvious and, not surprisingly, many have focused specific attention on resources utilized in food production, the efficiency with which these resources are used and the role animals play in this process (1-5). Efficiency of resource use in production, processing and storage of many food products, particularly those of animal origin, have been questioned and a great deal of publicity, often adverse to animal industries has been given to these issues. Statements made on both sides of the debate are summarized below.

- Large quantities of grain and other crop products directly edible by humans are currently fed to livestock.
- Because of their low efficiency animals reduce the energy and protein available for human consumption to a fraction of that contained in the feed.
- The production of edible energy and protein per unit land area is many times greater for crop production systems than for systems which include animals.
- As world demand for crop products for human consumption increases, amounts available for animal production will decline and future production of food by animals will become an immoral and unlikely possibility.
- Animal products are highly desired and of high quality, particularly in terms of protein and some vitamins and minerals.
- Productive quality of land varies considerably; over

half the world's land mass is non-arable and much of the arable land cannot sustain continuous, high yield cropping. Crop rotations including, for example, forage legumes are essential.

- Comparisons of animal and crop production systems often fail to consider differences in land quality.
- In most cases, less than one half of the dry matter produced by crops is consumed by humans; the remainder - field residues and processing byproducts - can be converted to human food by animals.
- Animals are able to harvest roughages from non-arable land and utilize inedible crop residues and byproducts to produce high quality human food.
- Domestic livestock represent a vast and highly flexible storage reservoir able to convert edible and inedible crop products in seasons of plentiful supply to high quality food in later times of shortage.

In aggregate, the general conclusion would seem to be that feeding of grains and other human edible crop products to livestock will decline in the future, at least in some areas, but that animals will continue to satisfy a significant portion of human dietary needs. Certainly it is clear that if food requirements of the world's expanding human population are to be met in the future, both in terms of quantity and quality, <u>all</u> available food production resources must be used effectively and efficiently. Unfortunately, these conclusions are entirely qualitative. This is a noticeable feature of many deliberations on this question; qualitative statements and data based on equivocal assumptions abound. The real challenge is quantitative consideration of the many elements and relationships involved (6). These include cultural, geographical, environmental and economic constraints as well as factors such as level of production technology, allocation of land and other resources, and alternative crop production and animal feeding and management strategies. Projection and planning of future development and integration of crop and animal agriculture will result in rational decision making only if these are based on rigorously derived and quantitatively defensable data.

It is our intention, in this paper, to assess briefly the current contribution of animals to the production of human food; to reconsider the concept and method of expression of efficiency and its implications on assessment of the future role of animals; and, to identify some of the critical issues which must be considered in the future. This in itself, will not provide the quantitative, systematic and inclusive description we advocated above; but to this objective we will present and briefly describe a model currently in development which may provide such an analysis of the human food production system.

The Present Contribution of Animals

The contribution of animal products to human energy and protein intakes for a sample of the 51 countries analysed in reference (6) are shown in figures 1 and 2. Total intake of energy and protein and the proportion supplied by animal products varies considerably. Countries represented on the left sides of the two histograms tend to be developed nations. Those on the right are less developed nations. As might be expected, a larger proportion of human dietary energy is supplied from plant sources; the average for all countries considered is 77.2 percent of energy from plant products and 22.8 percent from animal products. A major portion of protein intake particularly in the developed countries, is from animal products. Animal protein intakes in the less developed nations are lower and extremely variable. It is noticeable however that in at least a few developing countries, animal products provide a relatively high proportion of dietary protein, approaching or exceeding the overall mean of 46.6 percent.

As noted, animal products tend to contribute to a greater extent to dietary intake in developed countries. Latest available USDA statistics (7) on dietary intakes in the U.S. are shown in table 1. Approximately 35 percent of energy and 67 percent of protein is of animal origin. Also, animal products are a major source of some minerals and all vitamins excepting vitamin C. However, it is noteworthy that nearly half of our fat consumption is of plant origin.

An indication of the size of the animal industry in the U.S. is given in table 2. Farm - as opposed to retail - values of various segments of the industry are shown. In total, animal agriculture contributes over 41 billion dollars to the economy, or just over 4 percent of gross national product.

Clearly, animal agriculture contributes substantially to the economy and to total human food production worldwide. This, in itself, is a factor which should be given serious consideration in evaluating recommendations to reduce inputs to livestock production. However, the role of animals is not confined to the production of food. In developed industrial societies animal 'byproducts' supply a multitude of commodities for the pharmaceutical and cosmetic industries, industrial chemicals and feed ingredients as well as

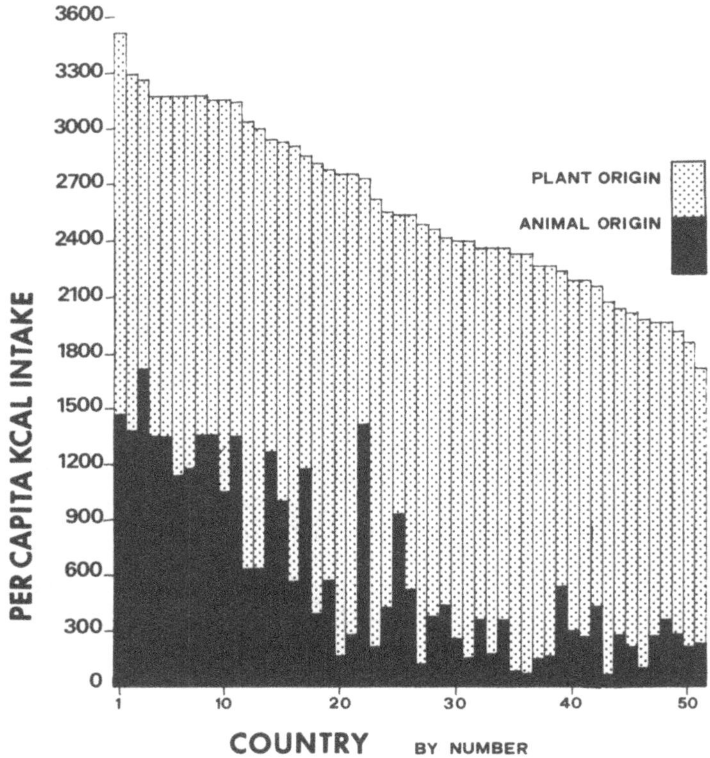

Figure 1. Intakes of plant and animal energy sources by the populations of 51 countries. Numerical code for each country follows: 1) Ireland; 2) USA; 3) France; 4) Austria; 5) Belgium; 6) Hungary; 7) Switzerland; 8) Federal Republic of Germany; 9) United Kingdom; 10) Argentina; 11) Australia; 12) Italy; 13) Israel; 14) Finland; 15) Norway; 16) Portugal; 17) Sweden; 18) Brazil; 19) Spain; 20) Egypt; 21) Turkey; 22) Uruguay; 23) Libya; 24) Chile; 25) Mongolia; 26) Paraguay; 27) Republic of Korea; 28) Japan; 29) Venezuela; 30) Pakistan; 31) Malawi; 32) Costa Rica; 33) Mauritius; 34) Panama; 35) Sri Lanka; 36) Burundi; 37) Nicaragua; 38) Nigeria; 39) Madagascar; 40) Honduras; 41) Kenya; 42) Colombia; 43) Ghana; 44) Philippines; 45) Guatemala; 46) India; 47) Ethiopia; 48) Ecuador; 49) Indonesia; 50) El Salvador; 51) Tanzania. From Baldwin, Slenning and Ronning, "A Visualization of the Livestock Industry in the World Perspective," in "Nutrition in Transition," Proceedings of the Western Hemisphere Nutrition Congress V, 1978. Reprinted with the permission of the American Medical Association and Publishing Science Group, Inc.

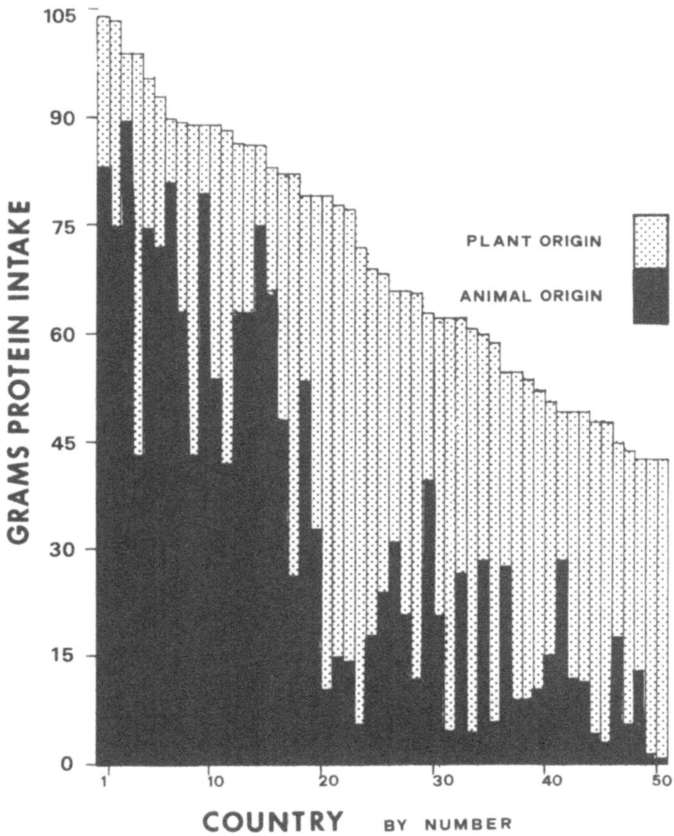

Figure 2. Intakes of plant and animal protein by the populations of 51 countries. Numerical code for each country follows: 1) Argentina; 2) France; 3) USA; 4) Hungary; 5) Ireland; 6) Mongolia; 7) Australia; 8) Belgium; 9) Israel; 10) Uruguay; 11) Finland; 12) Italy; 13) United Kingdom; 14) Austria; 15) Switzerland; 16) Federal Republic of Germany; 17) Norway; 18) Portugal; 19) Sweden; 20) Spain; 21) Egypt; 22) Turkey; 23) Japan; 24) Republic of Korea; 25) Kenya; 26) Brazil; 27) Chile; 28) Libya; 29) Ethiopia; 30) Paraguay; 31) Nicaragua; 32) Malawi; 33) Costa Rica; 34) Burundi; 35) Venezuela; 36) Nigeria; 37) Panama; 38) Honduras; 39) Pakistan; 40) Philippines; 41) Madagascar; 42) Colombia; 43) Guatemala; 44) Mauritius; 45) India; 46) Sri Lanka; 47) Ecuador; 48) El Salvador; 49) Tanzania; 50) Ghana; 51) Indonesia. From Baldwin, Slenning and Ronning, "A Visualization of the Livestock Industry in the World Perspective," in "Nutrition in Transition," Proceedings of the Western Hemisphere Nutrition Congress V, 1978. Reprinted with the permission of the American Medical Association and Publishing Science Group, Inc.

Table 1. Contribution of Animal Products to U.S. Dietary Intake in 1976 (in percentages). (7)

	Energy	Protein	Fat	Carbo-hydrate	Calcium	Phosph-orus	Iron	Vit.A	Thiamin	Ribo-flavin	Niacin	Vit.B_{12}	Vit C
Meat, Poultry & fish	18.2	35.8	32.9	0.1	3.3	23.0	27.7	20.2	27.9	21.4	43.0	65.2	1.2
Eggs	2.6	6.8	4.0	0.1	2.8	6.7	7.1	7.3	2.9	6.5	0.2	11.3	0
Dairy Products	14.7	24.4	19.5	7.7	76.5	38.3	2.3	13.3	10.5	44.8	2.0	23.5	5.1
Total	35.5	67.0	56.4	7.9	82.6	68.0	37.1	40.8	41.3	72.7	45.2	100.0	6.3

Table 2. Farm Value of Animal Production 1976 (in millions of dollars). (7)

Cattle & calves	13,988	Dairy products	11,724
Hogs	7,879	Chickens & broilers	3,095
Sheep & lambs	314	Turkeys	823
Wool & Mohair	96	Eggs	3,151
		Total	$41,070

plant fertilizer in various forms including manures. In many societies the range of non food goods and services provided by animals is perhaps even broader than this - to the extent that food production may not be the 'primary' product. The special role of animals in less developed countries is discussed more fully in a subsequent paper in this volume. The contribution of animals to human welfare through these many and varied additional services is often extremely difficult to evaluate. There is no doubt, however, that this contribution is real and of considerable magnitude.

Efficiencies of Animals as Producers of Human Food

The main question regarding animal agriculture however, is its net contribution to the production of food. More specifically, concern centers on competition between man and animals for crop products - particularly grains - and the efficiency with which animals convert these products to meat, milk and eggs.

Grain Use in Animal Production

Use of grains in animal feeds in the U.S. is certainly substantial. In table 3, statistics on grain production and animal feed usage are presented. These data are for 1975 which is the last year for which complete statistics are available (7). These data confirm that a large portion

Table 3. U.S. Grain Production and Animal Feed Usage in 1975 (in million tons). (7)

	Feed Grains			Food Grains			
	Corn	Sorghum	Oats & Barley	Wheat	Rice	Rye	Total
Produced	161.3	20.8	19.5	57.2	4.8	.6	264.2
Used as feed	99.6	14.3	13.2	1.6	0	.2	128.9
Percent feed	61.7	68.8	67.7	2.3	0.0	33.3	48.8

	Total livestock feed				
	Grains	Byproducts*	Harvested roughage	Pasture	Total
Million tons	128.9	36.6	89	243	497.5
Percent total feed	25.9	7.3	17.9	48.8	100.0

*Includes animal byproducts, milling byproducts and oil meals.

-close to half - of total grain produced is fed to animals. However, these grains represent only one quarter of total livestock feed; the remainder is roughages and byproducts which are inedible by humans. Also, as might be expected, almost all (98%) grains fed are feed grains, grain varities not generally consumed by humans. Nevertheless, it might be argued that land currently growing feed grains could be devoted to varities or species (e.g. wheat) that are consumed by humans. This apparently simple suggestion hides a number of very complex issues including input requirements and yields (and therefore, production costs and returns) of various grains in different localities (8); and, political, economic and social constraints to international distribution of food grains (9).

Efficiencies of Animal Production

Although these issues are complex and of real importance, the central issue remains the efficiency with which animals produce food. As already stated, a great many qualitative statements and questionable data have been promoted with respect to animal efficiency. A fundamental issue is the terms in which efficiency is expressed. If the ultimate concern is total human food production, it seems only rational to calculate animal efficiency in terms of inputs of human edible energy and protein and not in terms of total feed inputs. There is clearly no loss of human food and no competition for food which is not edible by humans in the first place. Total and human edible inputs and returns for a sample of livestock strategies are shown in table 4. These calculations are based on production

Table 4. Inputs and Returns from Animal Production.*

	Total energy & protein				Human edible energy & protein			
	Input		Return		Input		Return	
	E (Mcals)	P (kg)	E (%)	P (%)	E (Mcals)	P (kg)	E (%)	P (%)
Milk	19960	702	23.1	28.8	4555	111.5	101.1	181.4
Beef	20560	823	5.2	5.3	1869	39.9	57.1	108.8
Swine	1471	66	23.2	37.8	588	29.0	58.0	86.0
Poultry	23.2	1.2	15.0	30.0	11.2	.48	31.0	75.0

*Inputs are calculated as digestible energy and digestible protien and include cost of maintaining breeding herds and flocks.

practices common in California. Inputs include an allowance for maintenance of breeding herds and flocks. Rations were formulated by least cost methods used commercially. Taking the beef strategy as an example, for the production of one slaughter animal, total inputs amount to 20,560 mcals of digestible energy (DE) and 823 kg of digestible protein (DP). 5.2 percent of this energy and 5.3 percent of protein is returned in edible portions of the carcass. This is equivalent to a feed conversion of 19 to 1 for the total system (feed conversion of the animal from birth to slaughter is approximately 9.5 to 1). Though better than many publicised figures, this estimate is within the range of popular concepts of beef production efficiency. However, human edible inputs of DE and DP in the beef strategy presented amount to only 1,869 mcals and 39.9 kg, respectively or 9.1 and 4.9 percent of total inputs. Returns of these human edible inputs are 57.1 percent (1.75 to 1) for DE and 108.8 percent (.92 to 1) for DP. This particular system returns more human edible protein than is consumed. In the case of milk production, returns of both protein and energy are greater than 100 percent.

As might be expected, returns of human edible inputs from ruminants are higher than for non-ruminants because of the ability of ruminants to utilize a higher proportion of low quality, human inedible feeds in the ration. Even so, human edible returns from swine and poultry are higher than feed conversion figures normally considered. These data do not account for differences in quality between inputs and outputs. Since reductions of high energy feeding to livestock inevitably reduce animal performance, the possibility that some livestock production systems return more human edible energy and protein than they consume puts an entirely different complexion on questions concerning future roles of animals in human food production.

Protein returns for strategies considered in table 4 are consistently higher than returns of energy. This should not be surprising since livestock of all types are primarily producers of high quality protein, not energy. Many popular statements on animal efficiency use one conversion factor, often based on dry weight conversion but rarely stated as such, and apply it equally to all input and output components. Use of a single conversion factor whether adequately defined or not, has far more serious implications where it is implied, as has often been done, that the factor is some inherent biological constant applicable to or at least representative of all animals of the species and all production strategies. Such statements have tremendous public impact but are at best naive and at worst

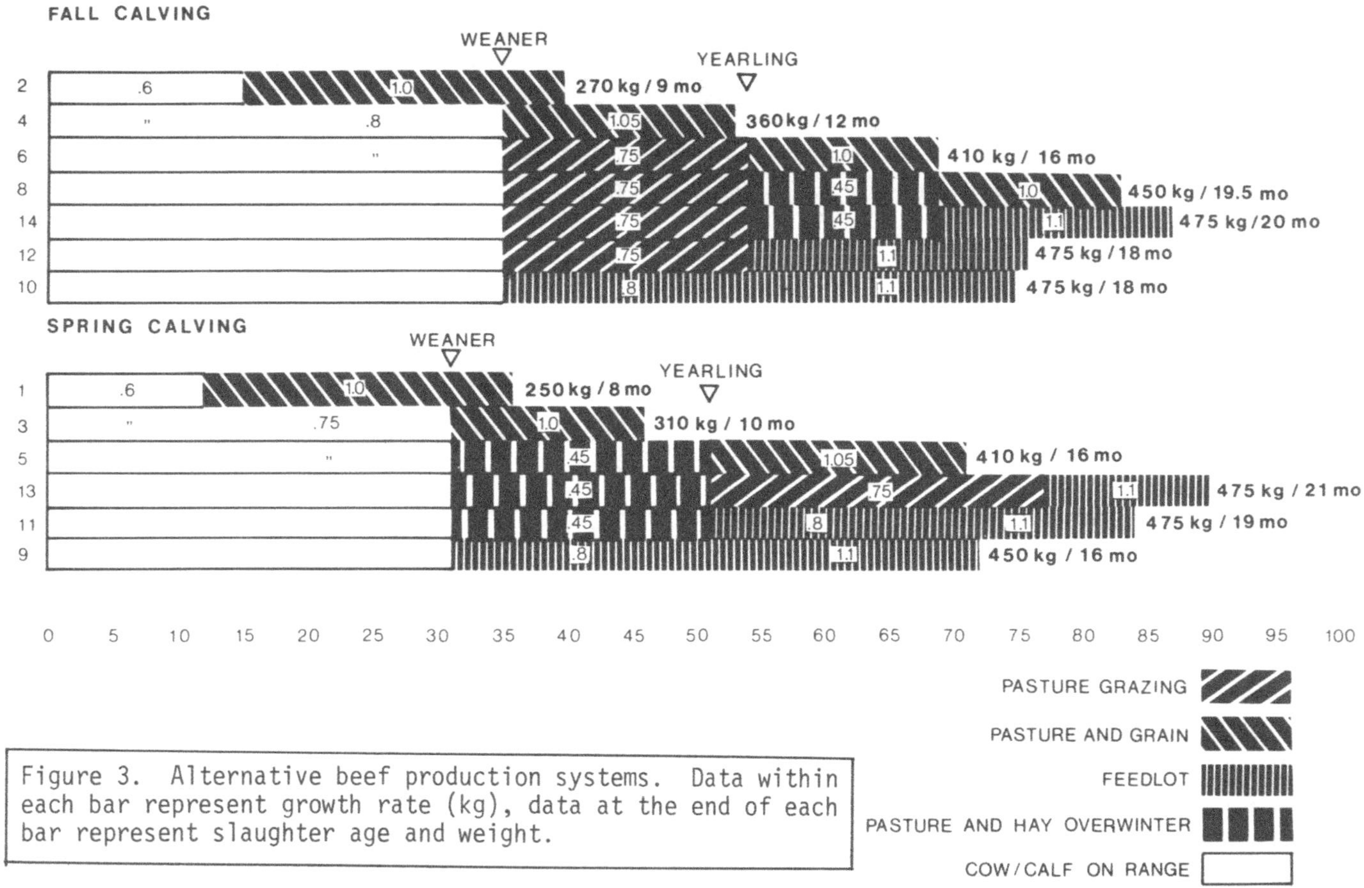

Figure 3. Alternative beef production systems. Data within each bar represent growth rate (kg), data at the end of each bar represent slaughter age and weight.

deliberately misleading. The data in table 4 are confined to one production strategy for each species for the particular purpose of emphasizing this point. It is as misleading to suggest that most beef animals under most production conditions return 108.8 percent of the human edible protein they consume as it is to suggest that most beef animals produce at a feed conversion efficiency of 5, 10, or 15 percent. There is no unique efficiency; every production system is different and, most importantly, varies over time dependent on feed resources available, economics and many other factors.

Alternative Strategies of Animal Production

The range of alternative strategies of livestock production is quite large and each particular strategy normally results in a unique production efficiency at a given point in time. Different strategies also require different quantities and types of input, produce different amounts of product of different quality, and may be more or less viable economically under circumstances of differing land, climate, feed availability, cost-price relationships and so on. An essential requirement for any analysis of future agricultural production is that such alternatives are included with respect to their impact on patterns of resource use and product output. Such a rationale is the basis of the model to be described in a later section of this paper. A number of alternative beef and dairy production strategies have been included in this model and some aspects of these alternatives will be considered now to elaborate upon the concept of variable animal efficiencies.

Beef

A partial list of beef production alternatives is illustrated in figure 3. Some of these strategies are duplicated at various stages allowing for utilization of different feedstuffs. The total number of strategies considered in the model is probably not inclusive in a nationwide sense - for the present the model is confined to California. However the range of strategies is sufficiently comprehensive in terms of input requirements and outputs to illustrate concepts considered here and to aid in identification of critical issues.

Three example systems have been chosen to illustrate methods of calculation (details of these calculations are given in the appendix). These are all Spring calving strategies with the calf weaned on range at approximately 180-205 kg (400-450 lb.) weight around the middle of

Table 5. Inputs and Returns for Three Beef Systems.

System	Input				Output		Returns			
	DE (Mcals)	DP (kg)	DE_H (Mcals)	DP_H (kg)	DE (Mcals)	DP (kg)	DE (%)	DP	DE_H	DP_H
5 Animal	6607	217.9	1234	28.1	658	30.5				
Herd	12072	435.6	114	2.6	146	8.1				
Total	18679	653.5	1348	30.7	804	38.6	4.3	5.9	59.6	125.7
11 Animal	9000	290.0	2815	61.1	921	35.3				
Herd	12072	435.6	114	2.6	146	8.1				
Total	21072	725.6	2929	63.7	1067	43.4	5.1	6.0	36.4	68.2
13 Animal	9512	318.0	1355	28.7	921	35.3				
Herd	12072	435.6	114	2.6	146	8.1				
Total	21584	753.6	1469	31.3	1067	43.4	4.9	5.8	72.6	138.7

[1]Systems are as described in the text; Details of inputs, output calculations are shown in the appendix.

[2]DE digestible energy; DP digestible protein; DE_H human edible DE; DP_H human edible DP.

September (7 months of age). The animal is over wintered on pasture plus hay and then either:

A) turned out on good grass with a self feed supplement of 90% grain and 10% salt - designed to limit intake to about 2.7 kg (6 lb. daily) - and slaughtered at the end of June at approximately 410 kg (900 lb.) at 16 months of age (system 5); or,
B) placed immediately in the feedlot, fed a growing ration to 315 kg (700 lb.) and finished on a high energy fattening ration to 475 kg (1050 lb.) at 19 months of age (system 11); or,
C) carried over spring and summer on pasture with no supplement, moved to the feedlot in late summer/early fall and finished to 475 kg at 21 months of age (system 13).

Estimated inputs to maintain a spring calving herd on range are 12,072 Mcals DE and 435.6 kg DP of which 114 Mcals and 2.6 kg respectively are human edible. Cull cows from the herd contribute 146 Mcals DE and 8.1 kg DP to total system output. Inputs and outputs to the animal and to the system in total with estimated efficiencies of return are shown in table 5.

At present all beef strategies considered in the model include some grain inputs during fattening. Much attention has been given recently to alternatives to the conventional high energy, feedlot systems for fattening beef. These include shorter periods in the feedlot, supplementation of pasture or harvested roughage with grain, and all grass finishing. Although comparisons between these experiments are sometimes difficult because of variations in rations, animal age, breed, growth rate and final weight, a number of points emerge fairly clearly. With the present market system and grading standards, some high energy input during fattening either as a supplement to grazing or in the feedlot is necessary to produce a carcass that will grade good or choice (10, 11, 12) and not be subject to a price penalty (12). Grain supplementation on grazing will generally produce an animal that is acceptable to the retail trade and the consumer (13, 14) though with a tendency to grade less consistently and slightly lower than feedlot animals (15). This is not to say that market conditions will not change. With an increasing proportion (currently around 45 percent) of beef being consumed in the U.S. as ground beef (16), there is the possibility that more distinct and more clearly defined market demands in the future will promote equally distinct production systems and marketing criteria. Such strategies will be investigated in

Figure 4. Energy output and efficiency of beef production.

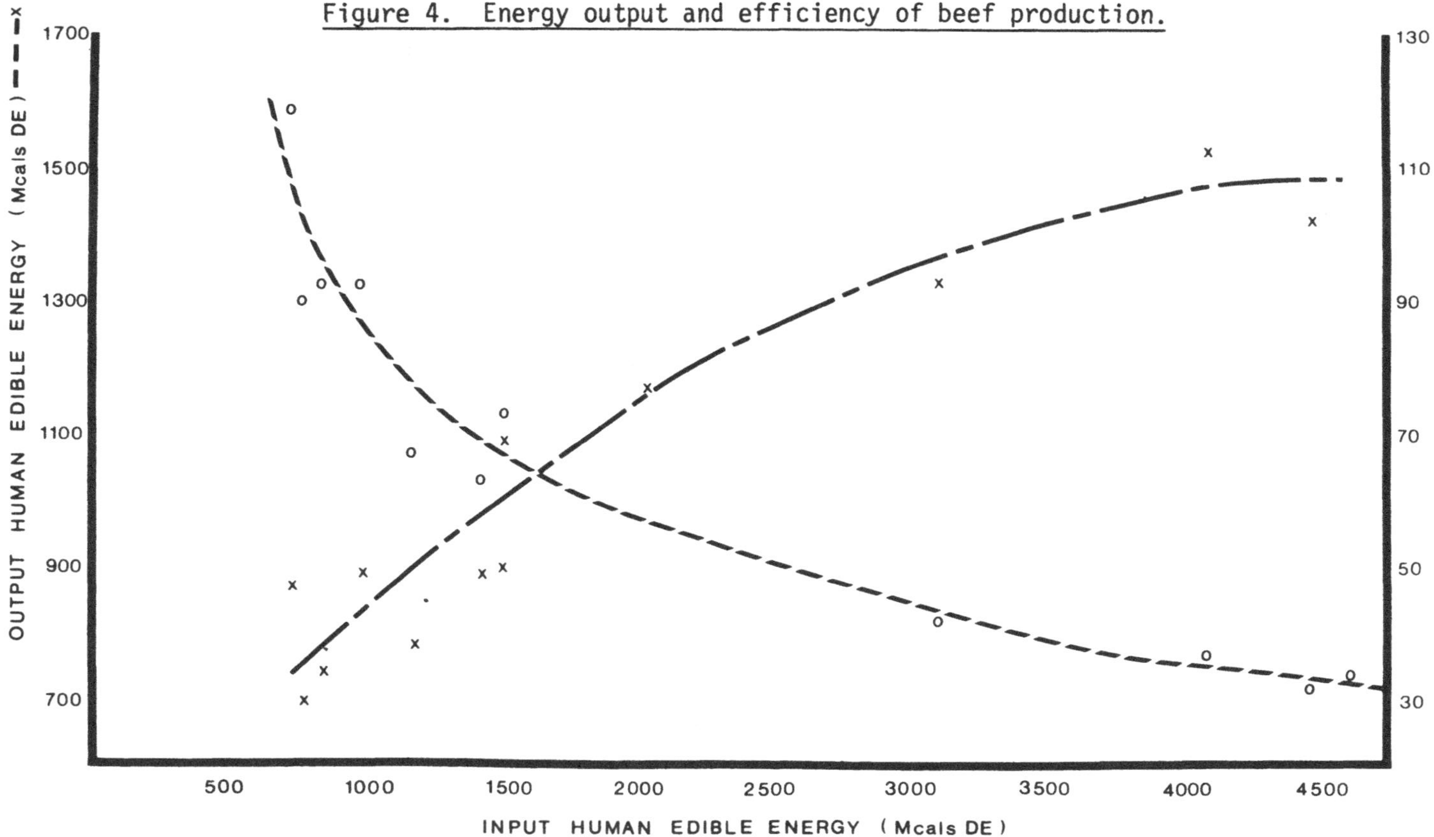

the model at a later date but for purposes of model development, strategies have been defined with respect to current market conditions.

Of more fundamental importance than market finish are the influences ration energy content, growth rate and slaughter weight have on various measures of efficiency for both the individual animal and the total system. The major determinant of total feed efficiency for the individual animal is growth rate which is determined largely by energy intake and ration energy content. Feedlot animals grow faster and convert feed to body weight more efficiently than animals supplemented on pasture which grow faster and more efficiently than unsupplemented animals (15). The result is a shorter time period required to reach slaughter weight (or a larger carcass in the same time period) and a lower total feed input for the same output. This is illustrated by comparison of systems 11 and 13 in table 5. A higher animal efficiency is reflected in a higher total system efficiency for system 11 through the large input to the breeding herd tends to reduce the difference. Since increases in growth rate and feed efficiency are achieved by increasing the portion of feed that is human edible, efficiency of return of human edible energy and protein for both the animal and the total system are markedly lower in system 11.

Effects of slaughter weight on efficiency can be seen by reference to data for system 5. There is a tendency, at least in California, for animals that are to be finished on grass plus grain to be fattened and slaughtered at an earlier age - although studies have shown that older animals (2 years plus) will fatten adequately and grade at an acceptable level on grass (15). System 5 results in a smaller carcass so that the large 'overhead' cost of the herd is spread over less output and total system efficiency is reduced. Human edible efficiency is lower than system 13 although human edible inputs are quite similar.

Taken together these factors represent the major variables not only in beef production but in most animal production systems. Reductions in feeding of high energy rations result in a decline in total output or a decline in total feed efficiency, or both, while increased feeding of high energy rations results in a decline in efficiency of return of human edible inputs. This can be seen quite clearly in figures 4 and 5 which show total output and percent human edible return against human edible input for all beef systems shown in figure 3, calculated as illustrated above. A fixed amount of grazing equal to that required by the system with the greatest requirement is assumed.

Figure 5. Protein output and efficiency of beef production.

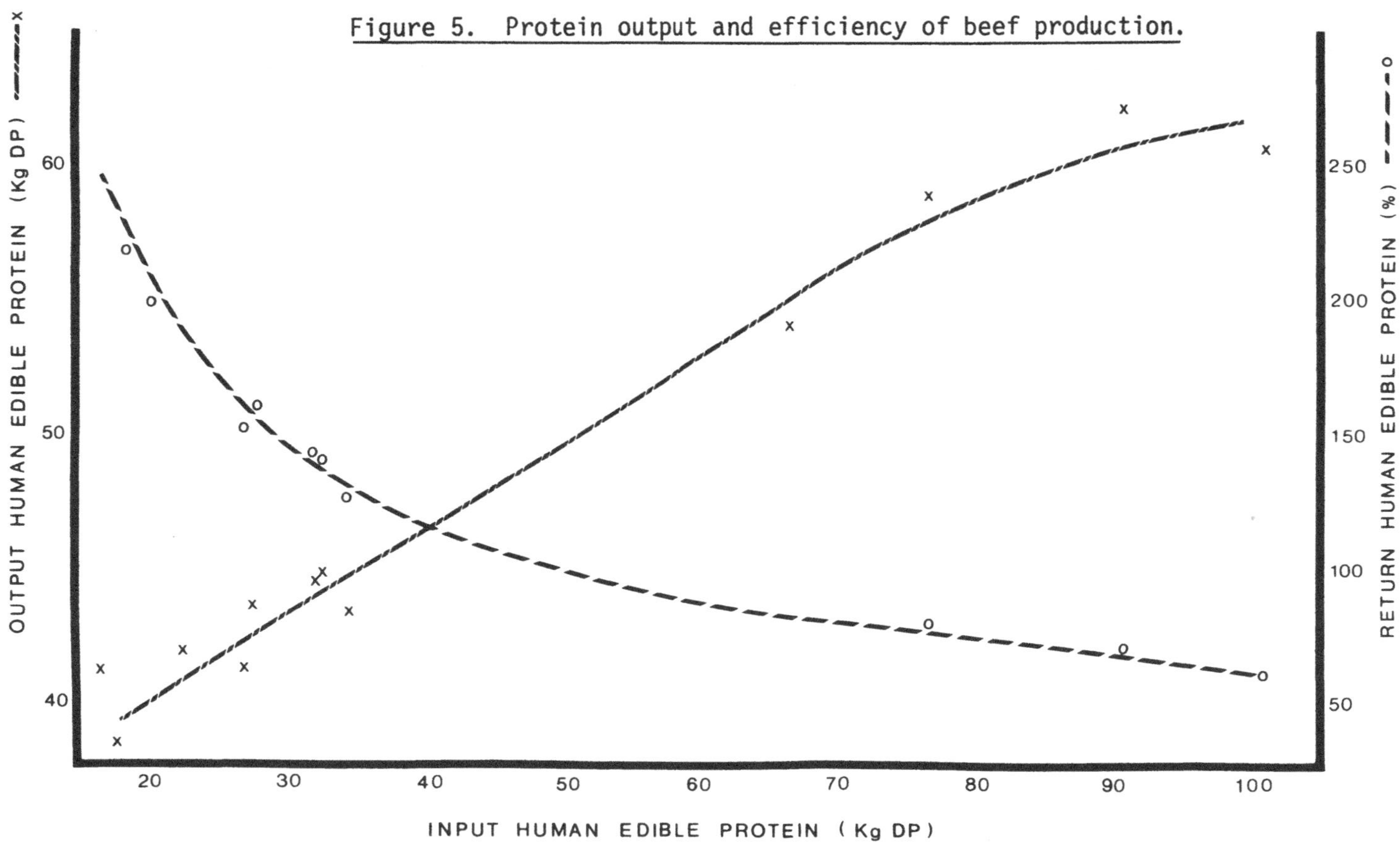

Total energy and protein output increase substantially as human edible input increases. At the same time, human edible return declines from just under 120 percent to under 40 percent for energy (figure 4) and from approximately 245 percent to around 60 percent for protein (figure 5). It is worth noting again that about half of the strategies considered return over 75 percent of human edible energy and all but four strategies return over 100 percent of human edible protein. These efficiencies are considerably different from those generally presented to the public. The real issue is that three quarters of the feed fed to livestock and closer to 85 percent of that fed to ruminants (17) is not edible by humans. This includes forages produced on untillable land such as the rangelands of the Western states. In California as in most of the region, range provides virtually all of the maintenance of the breeding herd and at least a portion of the feed input to slaughter animals. It does not provide , and because of relatively low and extremely variable quality, it cannot provide very much more than maintenance requirements. Of itself, range forage is not capable of promoting high levels of animal performance consistently and in adequate amounts. Thus, in order to realize sufficient quantities of useable product from our range resources, additional food sources must be utilized to captalize on the maintenance provided. Failure to do so represents a grossly inefficient use of what is a substantial proportion of total human food production resources.

Dairy

The same arguement applies equally to use of crop residues and byproducts. These are of greater importance in dairy production since, as a general rule, range grazing is not a viable option for the dairy industry. Figures 6 and 7 present data analagous to that in figures 4 and 5 but for dairy herds producing 3,175 kg (7,000 lb.), 5,000 kg (11,000 lb.), 6,350 kg (14,000 lb.), 7,700 kg (17,000 lb.) and 9,525 kg (21,000 lb.) of milk per cow per lactation. These levels of production are defined in the model to accommodate herd milk production averages achieved in California. The state average lactation yield is slightly over 6,350 kg (14,000 lb.) milk. A calculation of inputs and returns for this level of production is given in the appendix. Returns of human edible inputs are noticeably higher than for beef; the least efficient system returns about 94 percent of human edible energy and over 170 percent of human edible protein. Human edible output increases almost linearly with human edible input. Efficiency of total feed conversion is also shown in these figures. This increases in a very similar

Figure 6. Energy output and efficiencies of dairy production.

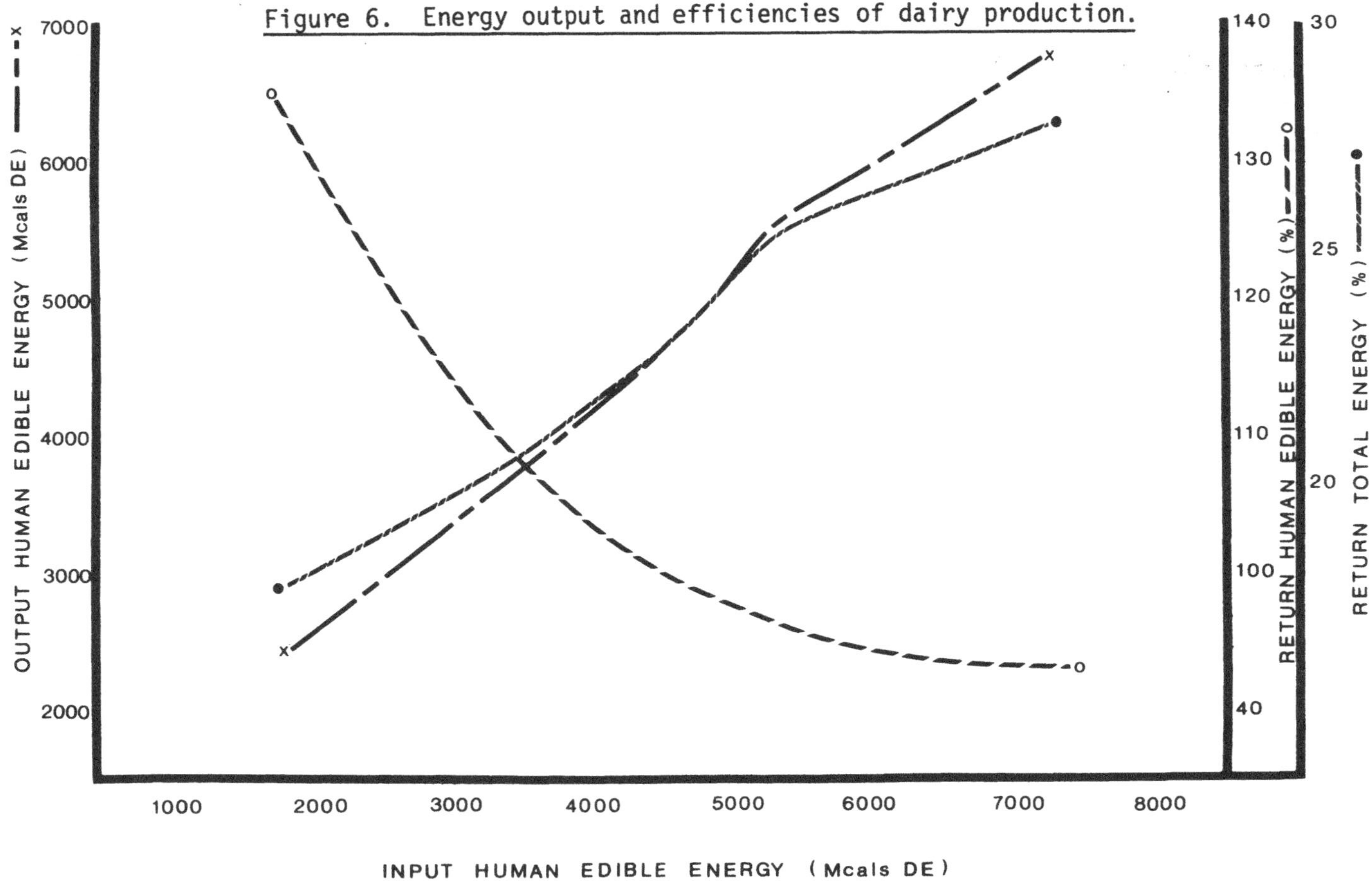

Figure 7. Protein output and efficiencies of dairy production.

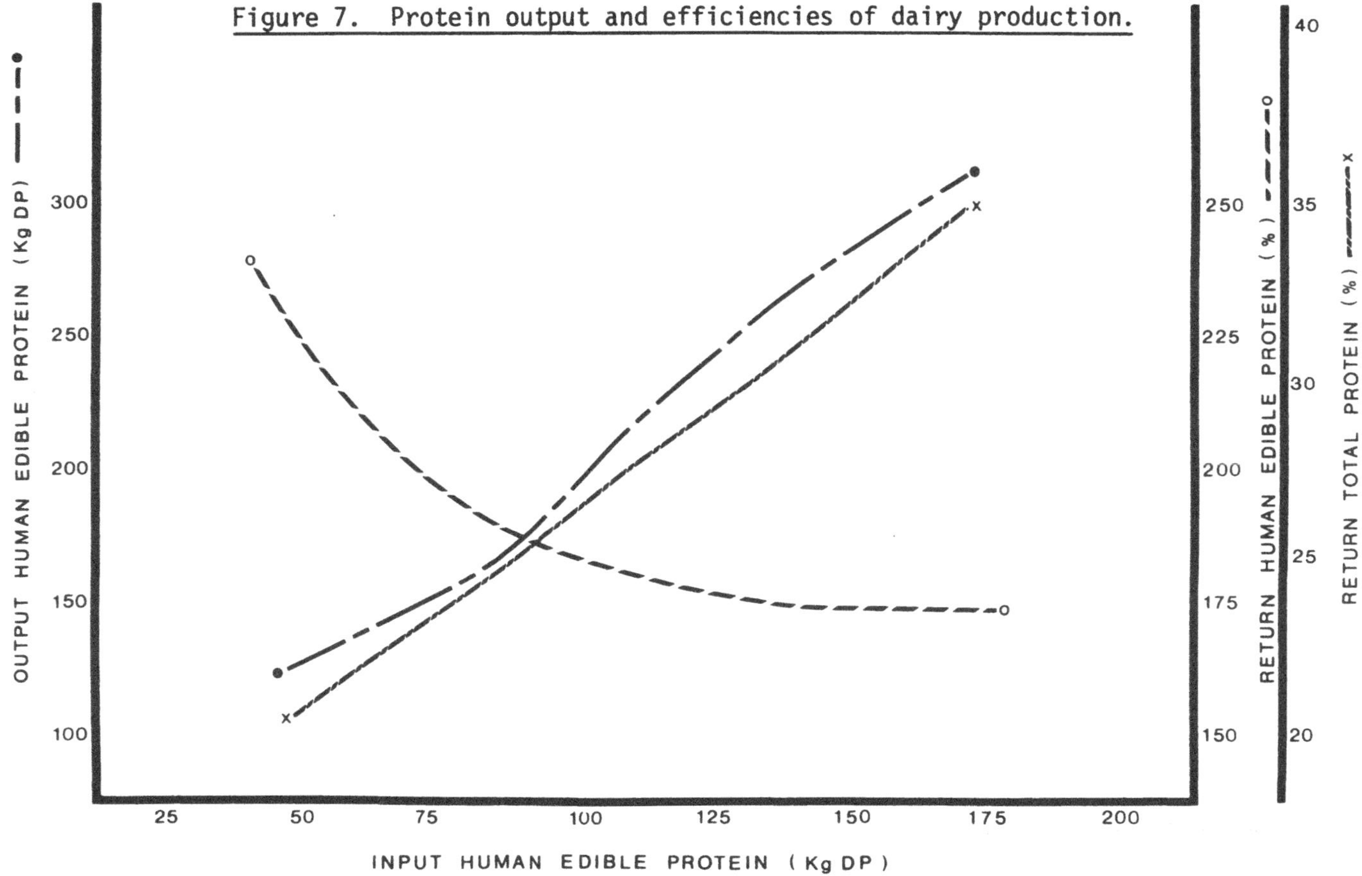

manner to total output -emphasising the point made earlier with reference to beef production. The inverse relationship between total feed efficiency and human edible efficiency is also clearly demonstrated in figures 6 and 7.

Economic Considerations

One factor which is often overlooked in this context is the question of production economics. To begin with, grain usage by livestock industries is primarily governed by price. This was amply demonstrated during the price increases of 1973 and 1974. The price of grains however is not influenced by the fact that they are human edible; it is influenced mainly by domestic supply and external demand. Economic efficiency from the producer's standpoint therefore bears little relation to the efficiency of human edible returns from production. Since feed costs are the major portion of variable costs however, economic efficiency is quite closely related to total feed efficiency, though that relationship is not perfect. If total feed efficiency declines because of a reduction of high energy inputs, economic efficiency might be expected to decline as well. Also, most intensive livestock industries such as dairies and beef feedlots require very high capital or fixed cost inputs. Spreading that cost over less output would aggravate the effect of declining efficiency. A combination of reduced total supply (output) of animal products and an increase in unit production costs would jeopardize the economic viability of individual producers. As evidenced by the events of 1973/74, this would inevitably result in significant increases in price.

The situation is complicated by the fact that byproduct usage in dairy and feedlot rations relies to some extent on the ability to combine them with high energy feeds. Byproducts by their nature are residues remaining after some product has been removed for human consumption. They are often of low quality and exhibit a marketed imbalance of major nutrients. Reduction in concentrate feeding and a greater reliance on lower quality but nutritionally complete diets would decrease usage of some byproduct feeds. Disposal of these by some other means would then have to be sought and approved to EPA standards. This would occur at an additional cost to the processor and ultimately to the consumer of the crop product.

While these latter comments are to some extent speculative, they indicate some of the complexity of interelationships between crop and animal agriculture and some of the issues to be faced if inputs to animal agriculture are to be reduced. It seems clear that such a move would be disruptive, not only to animal industries but to the total human food production sector. The only reason to contemplate such a move remains the arguement that animals are grossly inefficient as producers of human food. We have tried to show that this is not the case. We have also tried to show that a major portion of total resources available for human food production is not edible by humans initially and require processing through animals. If these resources are to be used efficiently they must be combined with some inputs of human edible foods. In short maximization of total human food production from all available resources may not only require maintenance of a viable animal industry, it may depend on diversion of a portion of human edible crop products to animal feed.

A Model of the Food Production System

Unfortunately the above conclusion is qualitative. The task of quantitative definition of actual and potential feed and food resources, their use by animals, and the economic viability, at least in the forseeable future of alternative crop production and animal feeding and management strategies remains. It is our contention that requirements for data relevant to analyses of future developments and integration of crop and animal agriculture can be accurately assessed and the data collected and analysed in a rigorous fashion. With these premises in mind we have attempted to formulate a model which could provide a quantitative, systematic and inclusive overview of the human food production system and the resources on which it is based.

Development of an appropriate and efficient method of describing and manipulating components of the food production system should provide a mechanism with which to explore alternative pathways and combinations of pathways of food production with as much objectivity as national or regional statistics allow. Also likely impacts of policy decisions or changes in economic or technological environments on an existing system and its resource inputs and food outputs could be quantified. The model is therefore, an aid to decision making in the agricultural sector and might have potential application not only in research but in public decision making as well. In the short run, the model might serve also as a public information tool to provide and demonstrate some of the hard data so often lacking in the past.

An initial objective is development and evaluation of methodologies appropriate to the type of analysis described.

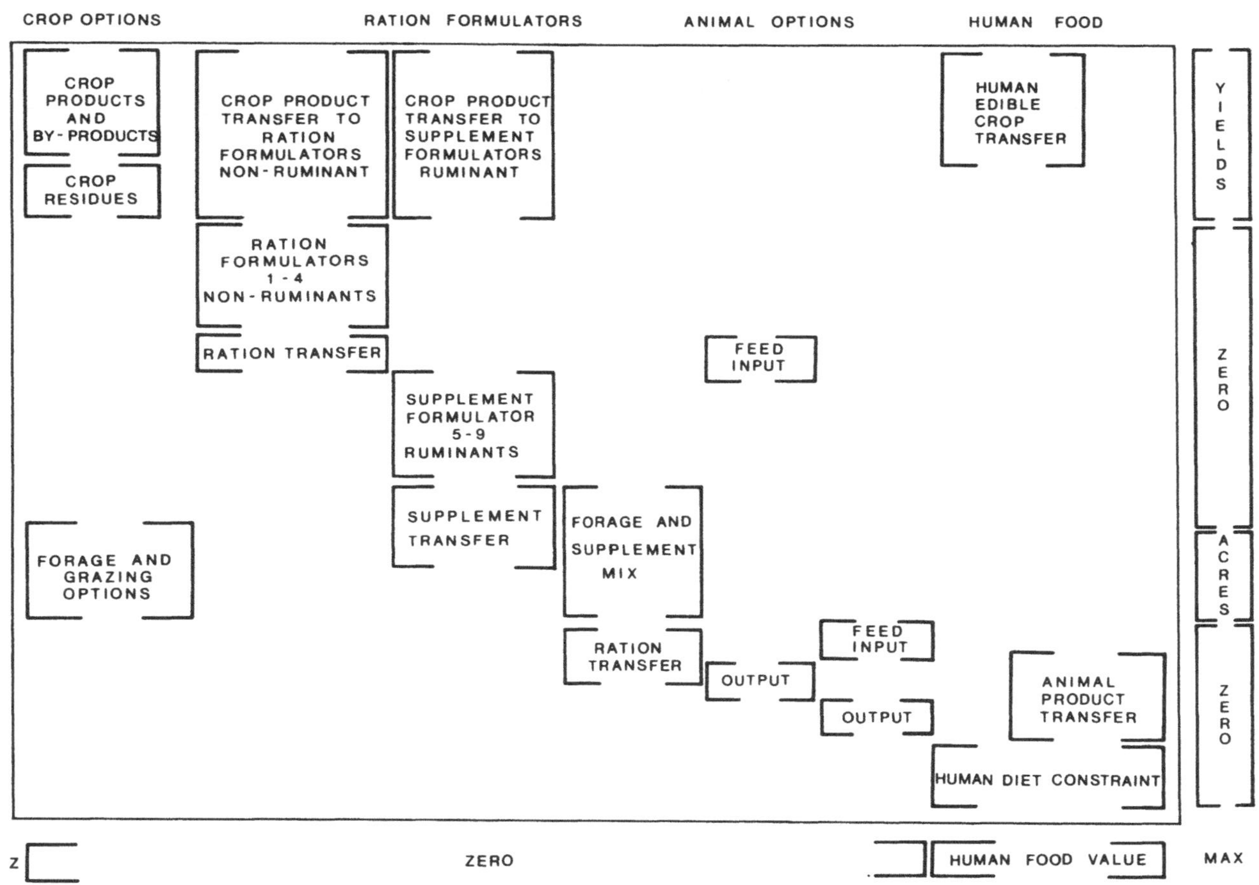

Figure 8. Generalization of the linear program matrix.

As a starting point, data have been collected for California and specified in the form of a linear programming matrix. The model explicitly represents a wide range of crops and the human edible and inedible products of processing; it includes a number of forage options and a range of alternative production strategies for the main animal classes with specific representation of nutrient requirements, feed formulation and production efficiencies; and, it provides for combination of various animal and crop products into a final human diet to provide minimum levels of required nutrients. A generalization of the matrix is shown in figure 8. On the left side of the matrix, the crop, crop products and grazing options currently incorporated include approximately 170 different commodities. This is a reflection of the great diversity of crop production in California. The central element in the matrix is comprised of 9 ration formulators. Four of these provide complete rations for hogs and poultry and five provide high energy supplements for ruminant livestock. These supplements are combined with roughages and grazing options in the ration blocks to provide complete diets appropriate to different classes of ruminants. The ration formulators are defined in terms of metabolizable energy and are structured in a format similar to least cost ration programs used in practice. As a result, rations allocated to various livestock in the model reflect rations which would be fed in practice, since they are generated in the same way. Total rations thus formulated promote production levels defined in an adjacent section of the model. In the past it has been limited consideration of these aspects, in particular, that has constrained the usefulness of speculations on future contributions of plant and animal agriculture.

Animal management options included in the model are all the beef alternatives described earlier, somewhat fewer sheep and dairy options, and fewer options for swine and poultry. Reduced options reflect constraints with these latter species in types of rations utilized. Due mainly to the large space requirement for each of the 9 main ration formulators, the matrix occupies, in total, 1700 columns and 400 rows. Not included in this number are certain elements which have been omitted from the first approximation in order to facilitate development. These elements, which will have to be accommodated in subsequent models include: specific allocation of land to each production alternative -the present representation assumes a fixed set of crop products and grazing options; monetary costs and returns from each production alternative - the model is confined presently to biological conversions only; and estimates of resource requirements other than feed, including for example, energy and water requirements. Expansion of the data set to include these elements will result in an extremely large matrix with, possibly, prohibitive computational requirements and costs. For this reason and in the interest of methodological comparisons it is intended to investigate the feasability of utilizing other techniques such as non optimizing methods to resolve the model.

Conclusions

Decisions concerning future developments of food production and the appropriate roles of animals and crops are clearly critical to human prosperity and welfare. The information base used in those decisions will determine their effectiveness to a very large extent and it should be a matter of great concern that current attitudes and awareness of these issues are apparently based largely on qualitative and for the most part very simplistic analyses of future options. The model described here is an attempt to provide the type of framework and quantitative information which represent the true state of our knowledge regarding the conversion of available resources into food.

APPENDIX

Calculation of Inputs and Outputs for Beef and Dairy Production

Beef Production
(calculation of data for table 5)

Herd inputs

Lactating Cow (450 kg):

February to June: heavy lactation, spring range.		
120 days @ 28.3 Mcals DE	3395	
June to September: light lactation, summer range + .7 kg/day supplement		
95 days @ 22.1	2097	
Including supplement		66.5 kg.
September to February: pregnancy, winter range + .7 kg/day supplement		
60 days, mid pregnancy @ 15.9	951	
90 days, last trimester @ 18.8	1690	
Including supplement		105 kg
Total DE requirement	8133 Mcals.	
Including supplement		171.5 kg

Dry Cows:

February to November: dry, pregnant from May, Spring/Summer Range		
185 days @ 15.7 Mcals per day	2905	
90 days @ 15.9	1431	
November to February: last trimester, Winter range + .7 kg/day supplement		
90 days @ 18.8	1690	
Including supplement		63 kg
Total DE requirement	6026	
Including supplement		63 kg

Heifers:

Birth to weaning:

February (born) to May: no grazing requirement, 32 to 80 kg		
May to September (weaning): 140 days .7 kg/day gain 80 to 178 kg		
80 - 125 kg 64 days @ 9.15	585	
125 - 178 76 12.7	964	
Less milk from cow.		
7.5 kg/day 140 days @ .688 Mcals/kg	-722	

Weaning to 2 years

September to following November: bred in May, 420 days, .5 kg/day gain, 178 to 388 kg. Range + summer & winter supplements		
178 - 225 kg 94 days @ 15.48	1455	
225 - 275 100 17.32	1732	
275 - 325 100 19.88	1988	
325 - 388 126 22.32	2812	
Including 1st winter suppl. 1.4 kg, 100 days		140
2nd summer 1.4 kg 135 days		189
November to February: last trimester pregnancy, range + 1.8 kg supplement		
84 days @ 22.56	1895	
Including supplement 100 days		180
Total DE requirement	10709	
Including supplement		509 kg

Bulls:

Assume 800 kg bull, roughage only (range, hay)
365 days @ 25.6
Total DE requirement 9348

Total Herd Input:

Calving rate 89% Replacement rate 27%
Breeding, 3 bulls per 100 cows.

	DE	Kg supplement
1 lactating cow	8133	171.5
.124 dry cow	744	7.8
.033 bull	311	-
.27 replacement	2891	137.4
Total	12079	316.7

Supplements: assume 70% hay, 10% each cottonseed meal, barley, molasses

	DE	DP
hay 221.7 kg @ 2.4 Mcals DE, 12.0% DP	532	26.5
cottonseed meal 31.7 kg @ 3.35 DE, 35.0% DP	106	11.1
molasses 3.9 DE, 10.0% DP	124	3.2
barley 3.6 DE, 8.2% DP	114	2.6
Range supplies (assuming Av. 2.0 Mcals DE 7.0% DP)	11203	392.1
Total herd input	12079	435.6
Human edible	114	2.6

Herd Output:

Assume 2% death rate. (replacement 27%)
culls = 25%
1 cow 450 kg 55.% dressing, 65.5% lean, 14.% fat
247.5 kg carcass 161.1 kg lean 34.7 kg fat
@ 8.4 Mcals/kg fat, 1.8 Mcals/kg lean, 20% protein in lean.
Total energy 291.8 (lean) + 291.5 (fat) = 583.3 x 25% = 146 Mcals.
Total protein 32.4 kg x 25% = 8.1 kg.

Slaughter Animal

Calf to Yearling:

Born February, no requirement until May, 32 to 80 kg wt.
May to September: 140 days,.75 kg/day gain, 80 to 185 kg, weaned on range.

80 - 125 kg	60 days @ 8.84 Mcals		530
125 - 175	67	12.05	807
175 - 185	13	15.95	207
Less milk from cow.			
7.5 kg/day for 140 days @ .688 Mcals/kg			-722
September to February: 140 days, overwinter pasture & hay .5 kg/day 185 to 260 kg			
185 - 225 kg	75 days @ 14.76		1107
225 - 260	65	15.49	1007
	Total DE requirement		2936 Mcals.
Protein equivelent @ 2.0 Mcals & 7% DP			102.8 kg.

System 5.

February to July: Good Pasture plus 2.7 kg/day supplement (90% barley)
140 days, 1.05 kg/day, 260 to 410 kg.

260 - 275 kg	14 days @ 20.49 Mcals		287	
275 - 325	47	24.61	1157	
325 - 375	47	26.83	1261	
375 - 410	33	29.27	966	
	Total DE		3671	
Less supplement, 381 kg @ 90%, 3.6 Mcals DE, 8.2% DP			1234	28.1 kg DP
	DE from pasture		2437	
Protein equivalent @ 2.8 mcals DE, 10% DP				87.0
	Total requirement		6607	217.9
	Human edible		1234	28.1

Output:

410 kg 59.% dressing 63.% lean, 21.% fat.
241.9 kg carcass 15.24 kg lean 50.8 kg fat
Assume 10% of fat trimmed and not consumed (remaining trim consumed)
Energy (1.8 lean, 8.4 fat) lean 274.3 fat x 90% 384 Total 658.3 Mcals
Protein (20% lean) 30.5 Total 30.5 kg

System 11.

February to May: Feedlot growing ration 3.2 Mcals DE 12% DP 60% concentrate 40% roughage: concentrate is 3.6 DE 10% DP 56.9% & 42.7% human edible;* 81 days .8 kg/day 250 to 315 kg.

250 - 275 kg	31 days @ 18.05		560	
275 - 315	50	22.8	1140	
		Total DE	1700	
		Protein		63.8
		human edible	580	13.7

May to September: Fattening ration 90% concentrate 3.5 mcals DE 10% DP
145 days 1.1 kg/day 315 to 475 kg

315 - 325 kg	10 days @	24.89 mcals	249	
325 - 375	45	27.31	1229	
375 - 425	45	30.43	1369	
425 - 475	45	33.72	1517	
		Total DE	4364	
		Protein		123.4
		human edible	2235	47.4
		Total requirement	9000 DE	290.0 DP
		human edible	2815.	61.1

Output:

475 kg 61.% dressing 61.% lean 27.5% fat.
289.8 kg carcass 176.7 kg lean 72.7 kg fat
Energy lean 318.1 fat (x90%) 603 Total 921 Mcals
Protein 35.3 Total 35.3 kg

System 13.

February to August: Pasture, 182 days, .75 kg/day, 250 to 385 kg

250 - 275 kg	34 days @	18.11	616
275 - 325	67	20.73	1389

* Feedlot concentrate contains 55% grain plus byproducts.

325 - 375	67	23.29	1561	
375 - 385	14	25.98	364	
		Total DE	3930	
Protein equivelent @ 2.8 Mcals DE 10% DP				140.4

August to November: feedlot fattening ration, 82 day 1.1 kg/day 385 to 475 kg.

385 - 425 kg	36 days @	30.43	1095	
425 - 475	46	33.72	1551	
		Total DE	2646	
Protein equivelent @ 3.5 DE, 10% DP				74.8
		Human edible	1355	28.7
		Total requirement	9512 DE	318 DP
		human edible	1355	28.7

Output: as for System 11.

Dairy Production

Herd average lactation yield 6350 kg (14000 lb.) 3.5% FCM.

Rations:

Concentrate supplement 3.4 Mcals DE 12.0% DE 46.4%, 29.4% human edible

For milk yield > 29.5 kg (65 lb.) ration	90% concentrate	3.3 Mcals DE
20.5 - 29.5 (45-65 lb.)	70%	3.1
13.6 - 20.5 (30-45 lb.)	50%	2.95
< 13.6 (30 lb.)	35%	2.8
Dry cow ration	20%	2.7

Mean Yields:

For lactation yield of 6350 kg mean daily yield assumed as follows

yield > 29.5 kg:	9 weeks mean yield	29.75 kg
20.5 - 29.5	13	25.06
13.6 - 20.5	13	16.89
< 13.6	9	11.57
Dry period	8.	-

Requirements[1]:

Maintenance 550 kg lactating cow. 17.6 Mcals DE/day
Dry/pregnant cow 22.9 "
Milk production 1.34 Mcals DE/kg milk.

9 wks. @ 29.75: 9x7x(17.6 + 29.75 x 1.34) 3620 Mcals				
protein @ 3.5 Mcals DE, 12% DP			124.1 kg.	
human edible		1512		34.8
13 wks @ 25.06	4657			
protein @ 3.3 Mcals DE, 12% DP			169.4	
human edible		1513		37.1
13 wks @ 16.89	3661			
protein @ 2.95 Mcals DE, 12% DP			146.4	
human edible		849		21.9
9 wks @ 11.57	2086			
protein @ 2.8 Mcals DE, 12% DP			89.4	
human edible		339		9.2
8 wks dry period	1282			
protein @ 2.7 Mcals DE 12% DP			57.0	
human edible		119		3.3
Total requirement	15306 DE		5863 DP	
human edible		4332		106.3

Output:

6350 kg milk @ 12% DM. 5.73 Mcals DE 24.8% DP (dry weight basis)
or .688 Mcals DE 2.98% DP (fresh weight basis)

Energy 6350 x .688 = 4369 Mcals DE
Protein 6350 x .0298 1892 kg DP

Herd Allowance:

Assume bull costs negligeable (service by AI)
Replacement rate 35%

Heifer requirements

Birth to 15 weeks: milk (replacer) + hay + starter (.23 kg (1/2 lb) per day for 10 wks) 40 to 100 kg

40 - 55 kg	35 days @	2.6 Mcals	.12 kg DP	91 Mcals	4.2 kg DP
55 - 65	23	4.0	.145	92	3.4
65 - 85	27	6.6	.245	178	6.6
85 - 100	20	8.8	.26	176	5.2
		Total		537 Mcals DE	19.4 kg DP

Including supplement
.23 kg/day 70 days @ 90% Dry matter 70% barley (3.6 Mcals DE, 8.2% DP)
human edible 79 1.8

[1]Energy requirements, minimum energy concentration taken from NRC Nutrient Requirements of Dairy Cattle, 1978.

15 weeks to 92 weeks (3 months prepartum) pasture plus hay with supplements (1.1 kg (2.5 lb) concentrate) for 2 winters (100 days) .75 kg/day 100 to 500 kg

100 - 125 kg	33 days @	8.8 Mcals DE &	.26 kg DP.	290 Mcals	8.6 kg.
125 - 175	67	11.9	.295	797	19.8
175 - 225	67	15.0	.33	1005	22.1
225 - 275	67	17.6	.365	1179	24.5
275 - 325	67	19.8	.395	1327	26.5
325 - 375	67	21.6	.43	1447	28.8
375 - 425	67	22.9	.465	1534	31.2
425 - 475	67	25.4	.495	1568	33.2
475 - 500	40	23.4	.505	936	20.2
			Total	10083	214.9

Including concentrates 1.1 kg for 200 days. 220 kg @ 3.6 Mcals DE, 12% DP: 46.4% & 29.4% human edible 367 7.8

92 weeks to 105 weeks (parturition): dry/pregnant ration.

90 days @ 23.4 Mcals		2106	
protein			93.6
human edible	195		5.5
Total requirement	12726 Mcals DE		327.9 kg DP
human edible	641		15.1
Total herd inputs @ 35% replacement	4454		114.8
human edible	224		5.3

Output-cull cow

550 kg 55% dressing 65.5% lean, 14% fat.
302.5 kg carcass 198.1 kg lean 42.4 kg fat
Energy lean 356.6 fat 355.7 Total 713.2 Mcals
Protein lean 39.6 Total 39.6 kg
Death rate 2% (replacement 35%)
Culling rate 33% Output 235 Mcals DE, 13.1 kg DP

Returns:

	Inputs (Mcals, kg)				Outputs (Mcals, kg)		Returns (%)			
	DE	DP	DE_H	DP_H	DE	DP	DE	DP	DE_H	DP_H
cow	15306	586.3	4332	106.3	4369	189.2				
herd	4454	114.8	224	5.3	235	13.1				
Total	19760	701.1	4556	111.6	4604	202.3	23.3	28.9	101.1	181.3

References

1. N. P. Guirry, A Graphic Summary of World Agriculture, USDA -ERS Misc Publ. 705 (1964).

2. H. DeGraaf, The Importance of Animal Agriculture in Meeting Future World Food Needs, Univ. Illinois, College of Ag. Special Publ. 12 (1967).

3. K. L. Blaxter, in Man Food and Nutrition, (CRC Press Cleveland, 1973) p. 127.

4. R. E. McDowell, Animal Production in World Food Supplies, Cornell Univ., Cornell Intl. Ag. Mimeo 45 (1975).

5. F. M. Lappe, Diet for a Small Planet. (Balantine Books, New York, 1971).

6. R. L. Baldwin, B. D. Slenning and M. Ronning, in Nutrition in Transition, Proc. Western Hemisphere Nutr. Congr. V (1978) p. 379.

7. USDA, Agricultural Statistics 1977. (Washington 1977).

8. C. R. Taylor, P. J. van Blockland, E. R. Swanson and K. K. Frohberg, Two national spatial-equilibrium models of crop production, Univ. Illinois Dept. Ag. Econ./AES Publ. 147 (1977).

9. USDA - ESCS, International food policy issues; a proceedings, Foreign Ag. Econ. Rep. 143. (1978).

10. J. C. Carpenter, R. A. Klett, P. B. Brown and G. L. Robertson, Producing quality beef with grass and grain, Louisiana State Univ. AES Bull. 627 (1968).

11. R. Mills, E. L. Martin, D. C. Anderson and R. L. Preston, Proc. Western Sect. Am. Soc. Anim. Prod. 29 (1978) p. 94.

12. J. Shinn, C. Walsten, J. L. Clark, C. B. Thompson, H. B. Hedrick, W. C. Stringer, A. G. Matches and J. V. Rhodes, J. Anim. Sci. 42. 1367. (1976).

13. R. A. Bowling, G. C. Smith, Z. L. Carpenter, J. R. Dutson and W. M. Oliver, J. Anim. Sci. 45. 209 (1977).

14. J. C. Carpenter and P. B. Brown, Performance and feed costs for selected methods of feeding beef calves

from weaning to slaughter, Louisiana State Univ. AES Bull. 612 (1966)

15. R. A. Bowling, G. C. Smith, Z. L. Carpenter, R. L. Reddish and O. D. Bulter, J. Anim. Sci. 46. 333 (1978).

16. Economic Report. Supplement to the Washington Report by the National Restaurant Assoc. (July, 1978).

17. L. S. Pope and L. M. Schake, The Cattleman 62. 52 (1975).

Larry Martin, Karl D. Meilke

2. Implications of International Trade on Resources for Animal Production

During the decade from 1966 through 1975, U.S. feedgrain exports increased from less than 10 to over 20 percent of domestic production (see Figure 1). During 1974-75, the period when exports increased most rapidly, U.S. feedgrain prices were 230 percent higher than in 1971-72. Livestock production subsequently dropped markedly in response to these feedgrain prices.

Considerable debate has recently surrounded a number of potential policies affecting North American grain markets. The policy alternatives range from a multilateral "OPEC-like" grain cartel at one extreme to relatively unconstrained production and trade, aimed at maximizing foreign exchange earnings at the other.

Most of the debate and attendant analyses[1] of grain policy has emphasized the economic impacts of grain policy on only the grain sector. The contention of this paper is that policies affecting the domestic or international trade components of feedgrain markets have repercussions in the livestock sector which are of sufficient importance to be considered in policy decisions. These repercussions are important to society and, therefore, to policy makers for at least two reasons. First, grain policy affects production, consumption and prices of livestock and livestock products and therefore the producers and consumers of these products. Second, through their repercussions on livestock production, feedgrain policies indirectly affect the domestic economy by altering investment and employment patterns in manufacturing and tertiary industries connected with them: e.g., meat

The authors appreciate the valuable assistance given by Mike Farrow and Ellen Goddard during the preparation of this paper. This project was supported in part by the Ontario Ministry of Agriculture and Food and Agriculture Canada.

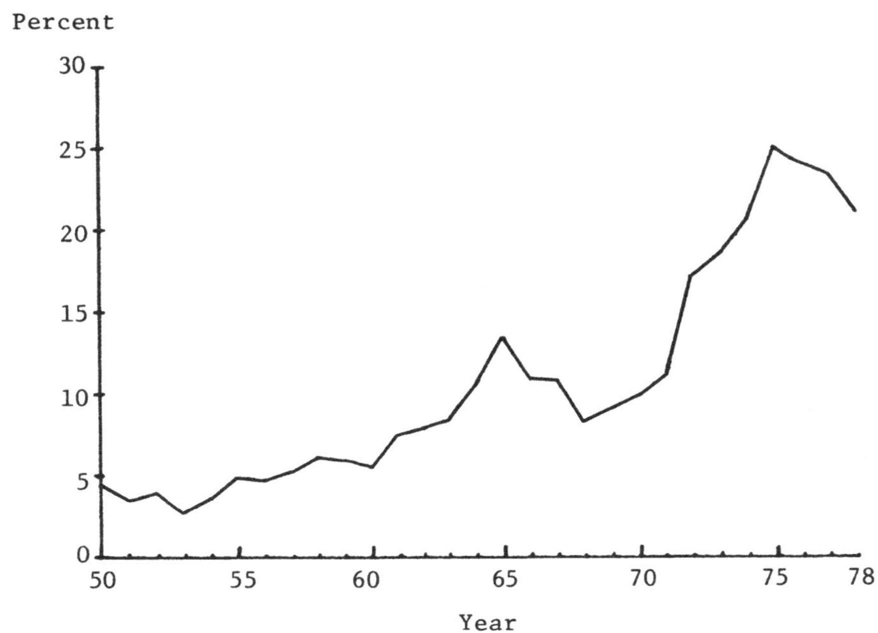

Figure 1. U.S. feedgrain exports as a percent of total supply.

processing, transportation and distribution.

One objective of this paper is to provide, using econometric techniques, preliminary indications of the magnitude of costs and benefits to the U.S. feedgrain and livestock sectors of alternative policies affecting the feedgrain sector. A second, and more basic objective, is to identify the kinds of information and interactions between the livestock and feedgrain sectors which are necessary for such analysis. The paper has two parts. In the first, our approach to modeling the feedgrain-livestock sector is discussed in order to illustrate the kinds of information and interrelationships needed to analyze policy situations. In the second, four policy scenarios are addressed.

Approach to the Analysis

Our original intent in carrying out the analysis was to construct a relatively small econometric model of the U.S. feedgrain, beef and pork sectors. This model was to have been a subset of a much larger North American agricultural sector model being constructed jointly by Agriculture Canada and personnel in five Canadian universities. The model contains quarterly behavioral relationships for: domestic supply, demand and carry over of feedgrains; export demand for feedgrains; domestic supply, demand and carry over for pork, grain-fed and grass-fed cattle (including cows and bulls); and exogenous import/export demands for pork and beef.

To complete the analysis, the intended approach was to combine the estimated equations into a simulation model which would generate the supply, demand, carry over and trade variables discussed above, as well as livestock and feedgrain prices from 1971 through 1985. Then the various policy scenarios were to be compared with a base run over this period. However, after considerable estimation, simulation and diagnosis, the livestock component of this model was found to have explosive properties.[2] In order to present some preliminary, empirical analyses, we adopted a modified approach. The feedgrain component of our original model was interfaced with an already existing, recursive spatial equilibrium model of the North American red meat sector. Policy simulations were run over the period 1970 to 1978.[3,4]

While this modified approach is similar to the original approach and, in general, includes the same behavioural relationships, it has limitations. The most obvious and most serious is that the grain and livestock sectors are not fully simultaneous.[5] As will be seen below, because of relatively long lags in the supply equations of the livestock model, the impacts of a change in grain policy during the early to mid-1970's begin to be significant to the livestock sector only in 1976 or 1977. Hence, the largest impacts would likely be subsequent to the latter date; but, because the two models are not fully simultaneous, it would not be credible to run the simulation over a longer time period. As a result, the analysis does not measure full impacts on the livestock sector.

A second problem is that there are a number of both subtle and not so subtle differences in the data and estimation procedures used to develop the two models. This means that the linkages between the two models are likely weaker than they would be if properly developed. A major example of these differences is that the recursive spatial equilibrium model does not separate grain and grass-fed beef as did our econometric model. This likely results in an understatement of the magnitude of the impact of a change in grain prices on the beef sector and a misstatement of the timing of those impacts.

The net effects of having to modify our approach to the analysis are likely:

1. The magnitude of intercommodity impacts may be underestimated.
2. Full impacts of a change in grain policy on the livestock sector are most certainly understated because of the limited simulation period.

However, the direction of change and the timing thereof indicated by the analysis illustrate the importance of understanding and properly quantifying interrelationships between the grain and livestock sectors.

Nature of Interrelationships Between the Feedgrain and Livestock Sectors

While the supply and demand for feedgrains, beef and pork are all related to their own prices (often lagged appropriately), each supply and demand is related to prices and quantities of other products. The most important interrelationships are in domestic demand for feedgrains and domestic supply of livestock.

Domestic Demand for Feedgrains

Domestic demand for feedgrains is obviously affected by the size of the domestic livestock herd. But, it is also affected by livestock prices. If livestock prices increase relative to grain prices, feedgrain demand may increase in the short run because livestock are fed to heavier weights and, for beef, a larger portion of the feeding age cattle are grain-fed rather than grass-fed. In the longer run, demand is affected because an increase in the relative prices of livestock provides an incentive to increase the size of the domestic breeding herd which, in turn, leads to an increase in demand for feedgrain. Naturally, the opposite effects occur if livestock prices fall relative to feedgrain prices.

The foregoing is the first round of effects. The relationship also has secondary effects. If livestock numbers or prices increase in the short run, thereby increasing the demand for feedgrains, this results in a short run increase in grain prices and, therefore, an incentive to increase grain production in the longer run.

Domestic Supply of Livestock[6]

Livestock supply is affected by feedgrain prices in various ways. Most obviously, since grain is a major input in the livestock production process, a relative increase in feedgrain prices increases production costs, and, eventually, reduces livestock production. But, since a large proportion of North American farms are integrated operations which produce both grains and livestock, in extreme circumstances feedgrains can become a competing marketing enterprise with livestock. This likely happened for a number of farms during the mid-1970's when the dramatic increase in grain prices made selling grain as grain more profitable than selling it as livestock. The result was liquidation of a portion of the breeding herds of, first, hogs and, subsequently, beef.

Whether feedgrains are regarded as an input or a competing enterprise for livestock, the generally expected relationship is that a relative increase in grain prices will result in a decline in livestock production and vice versa when relative grain prices fall. But the magnitude and, particularly, the timing of these impacts, are subtle and difficult to quantify as they filter through producers' decision processes. There are both short and longer run impacts.

In the short run, an increase in grain prices would be expected to result in hogs or cattle on feed being marketed at earlier ages and lighter weights. Similarly, an increase in grain prices could increase the age and weight at which range cattle are placed on feed or, at the extreme, result in marketing them directly off grass. Again, grass fed cattle would likely be marketed at lighter weights, thereby reducing beef supply. Furthermore, increased feed prices are bid rapidly into the prices of feeder cattle and feeder hogs. If the female animal is viewed as a capital investment, her value in the breeding herd is the discounted present value of the expected flow of receipts from the sale of her progeny. If an increase in grain prices results in lower feeder cattle or feeder hog prices, then the capital value of the female animal held for breeding falls. This reduces the incentive to hold or add females to the breeding herd; so more are slaughtered. Hence, in the short run, beef supply might increase because of female slaughter, but will decrease in the longer run because smaller breeding herds lead to smaller calf and pig crops.

The expected overall net effect of an increase in grain prices is an increase in livestock supply in the short run and a decrease in supply over the longer run. The longer run, negative, effect is expected to outweigh the short run, positive effect. More important, given the biological and decision lags required to change breeding decisions, complete the gestation period and rear progeny to market age, the long run is indeed long. In our livestock model, the major impact of a change in grain price on livestock supply is between 1½ to 2½ years for hogs and between 3 and 4½ years for cattle.

The length of these lags leads to some interesting and important implications for the policy making process. The lags suggest clearly that a policy decision which affects the current market situation for one commodity may have serious repercussions for another considerably later. This will be illustrated below in the policy analysis.

Policy Analysis

A wide range of policy instruments is available to government with which to affect the feedgrain/livestock sectors. Four sets of instruments are addressed herein: (1) fixed exchange rates; (2) an export tax on feedgrains; (3) a combination of no price freeze on meat, as in 1973, and a quota on feedgrain exports; and, (4) a buffer stock policy for feedgrains. The first three scenarios are analyzed with our econometric models. The fourth is addressed by referring to previous and more thorough research.

Model Validation

For analysis of the first three policy scenarios, the models of the feedgrain and livestock sectors will be used. It may, therefore, be of value to indicate briefly the base to which the policies will be compared and the ability of the models to simulate actual events in the markets. The models were validated from the fourth quarter of 1969 through the fourth quarter of 1977 for feedgrains and the second quarter of 1978 for livestock. Figure 2 contains actual and simulated values for U.S. corn prices, beef cow (breeding) inventories, beef and pork production, and steer and hog prices, respectively. All values except beef cow inventories are quarterly. Beef cow inventories represent observations taken at January 1 of each year. Hog prices are measured in dollars per cwt. on a carcass weight basis (as they are quoted in Canada). Live weight prices for hogs are approximately 75 percent of carcass weight prices.

As can be seen, the feedgrain model simulates corn prices relatively well. There are, however, two periods of error which are important for the livestock model. The model generated corn prices in 1971 and 1972 which were higher than actual and, for two quarters in 1976, model generated corn prices were nearly $.50/bu. lower than actual.

Model generated beef cow inventories are lower than actual in 1974 and 1975 and higher than the actual for 1978. Much of the error in 1974 and 1975 results from the overestimation of corn prices (and therefore the underestimation of feeder calf prices) in 1971 and 1972. As a result of these errors in cow inventories, beef supply during 1976 is underestimated. This resulted in higher than actual steer prices for 1976. This error, plus the error in 1976 corn prices, means that with the model, feeder calf prices for 1977 are substantially overestimated. This results, in turn, in overestimation of cow inventories in 1978. When actual grain prices are used as input to the livestock model, actual and model generated cow inventories are much closer.

Beef production and steer prices are reasonably well simulated by the model until 1976 when beef supplies are lower and steer prices higher than actual as discussed above. Then, in 1977 and 1978, model generated beef supplies are higher than actual because of the errors in steer and corn prices in 1976.

Pork production and prices are simulated reasonably well until 1977. From that point onward, the model substantially overestimates production and underestimates prices. These errors occur because the model does not account for the poor winter weather and disease conditions which affected the U.S. pork industry during 1977-78 and because there was a lag in investment in this industry, probably in response to the high grain prices of the mid-1970's, which the model would not

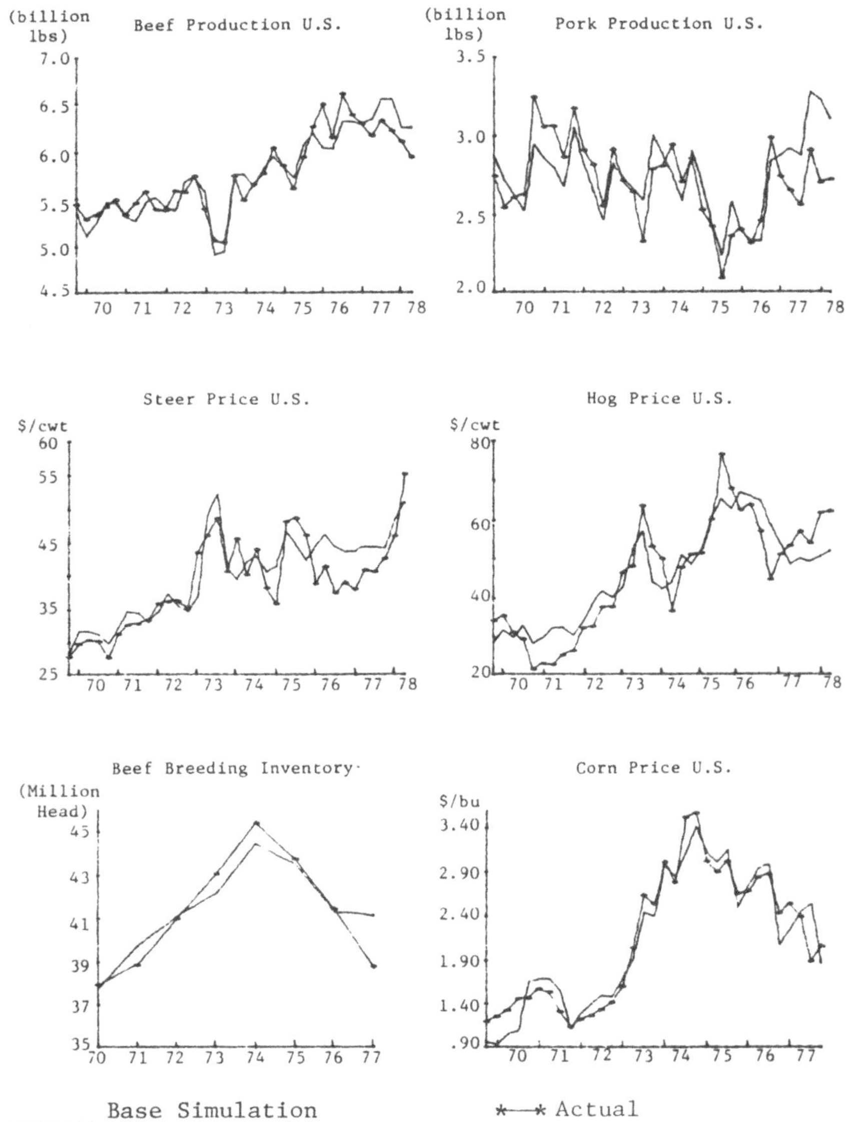

Figure 2. Actual and base simulation, U.S. 1969-4 - 1978-2.

have reproduced even without the excuse of poor weather and disease.

With the foregoing limitations, the models seem to represent the sectors relatively well and the validation run of the models will be used as the basis against which the following policies are analyzed.

Fixed Exchange Rate Policy

In 1971, the United States allowed the dollar to float after a long period of having fixed its value. As a result, the dollar has subsequently devalued. Had the dollar been maintained at its pre-1971 value, one would have expected U.S. feedgrain prices to be lower because the higher value of the U.S. dollar, relative to other currencies, would have decreased export demand for U.S. feedgrains.

In this analysis, the dollar's value was maintained at the pre-1971 level relative to Western European and Japanese currencies after 1971. As expected, this run resulted in lower domestic feedgrain prices (Figure 3), although it did little to stabilize them during 1974-75. The impacts of this policy on average quarterly corn prices, exports and total revenue from sales of feedgrains are presented in Table 1. Note that exports fall by nearly 15 percent, price is 4 percent lower, and revenue drops by 5.2 percent.

Table 1. Effects of Fixed Exchange Rates on U.S. Corn Prices, U.S. Feedgrain Exports and Total Revenue from Feedgrain Sales, 1971-1977 (Quarterly Average).

	Base Run	Fixed Exchange Rate	Difference
Corn Price ($/bu.)	$2.33	$2.22	$-0.11
Feedgrain Exports (m.m.t.)	9.81	8.36	-2.32
Total Revenue (mil. dol.)	$4,116	$3,892	$-224.1

The effect on U.S. livestock production and prices can be seen in Figure 3. Note that while the impact on beef production through 1978 was relatively small, the impact on breeding inventories becomes progressively larger through time. Breeding inventories would have been built even higher in 1975, would not have experienced as rapid a decline in 1976 and 1977, and would have increased subsequently. The larger (than base simulation) calf crops resulting with the

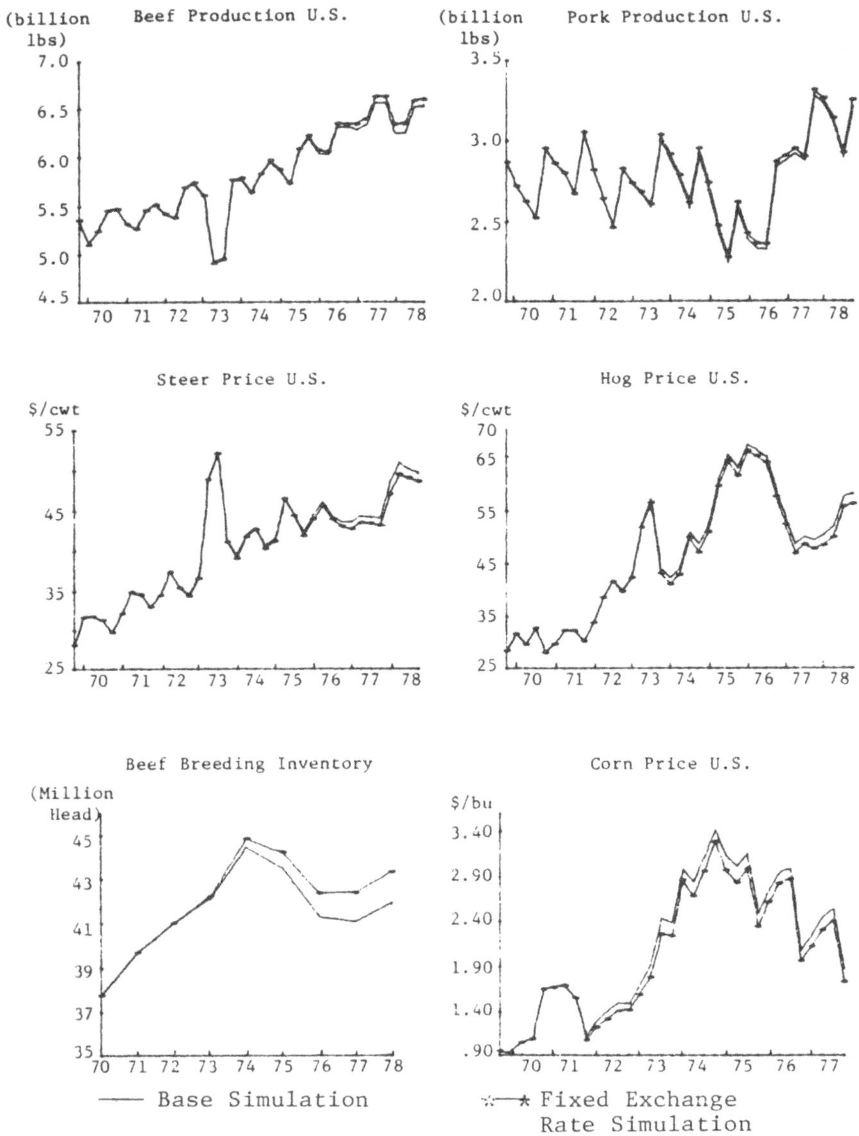

Figure 3. Fixed exchange rate and base simulation, U.S. 1969-4 - 1978-4.

exchange rate policy would lead to larger beef slaughter and lower beef and pork prices from 1979 through 1981. Hence, the major impact of the exchange rate policy would be realized after the simulation period.

To summarize the impacts of this policy on the livestock sector, a quarterly average net revenue to the feeding component is calculated. This is derived by subtracting the total cost of feed from the total revenue generated from livestock production. Total feed cost is, in turn, calculated by assuming that 800 lbs. of corn is required to produce 100 lbs. of live steer and 600 pounds of corn is required to produce 100 lbs. of live hog, (both of these factors include an adjustment for maintenance of the dam or sow). Using these assumptions, the total grain requirements for the amount of livestock can be estimated and cost of feed calculated for a given quarter by using the model generated corn price lagged two quarters (on the assumption that feed is purchased before livestock are marketed). Hence, by calculating total revenue and total feed cost for a given quarter, net revenue can be found by finding the difference. Then the average quarterly revenue for the base and policy can be calculated and compared.

There are, of course, many limitations to this measure of net benefit. Since only corn is used in the calculation, other feed costs (protein and pasture for grass fed cattle) are ignored as are other variable production costs. Furthermore, total revenue is calculated using only steer and market hog prices; no account is taken of the proportion of slaughter represented by heifers, cows, bulls or sows. Finally, to attribute all the benefits to the feeding component of the livestock sector is clearly wrong. Much, if not most, of the benefits would be passed on from feedlot operators to those who produce replacement cattle or hogs in the form of increased feeder cattle or feeder hog prices.

Nevertheless, since feedgrains represent a major production input, this measure provides at least rough justice to the measurement of benefits. Average quarterly net revenue for the base and fixed exchange rate situations are presented in Table 2. The net gain to the livestock sector of $121.04 million is less than the loss in revenue to the feedgrain sector of $224.1 million (Table 1). However, the loss in revenue to the feedgrain sector is an overstatement because no account is taken of production costs, which would be lower with the exchange rate policy because of lower grain production and potential input substitution. Similarly, the impacts on the livestock sector, particularly the beef component, are misstated since the major impact on beef supply would occur after the simulation period. It is not obvious whether the gains would be larger or smaller after 1978, since increased beef supply would result in lower cattle and hog prices which could offset lower feed costs.

Table 2. Effects of Fixed Exchange Rates on Net Revenue to the Livestock Sector, U.S., 1971-1977 (Quarterly Average).

	Base Run	Fixed Exchange Rate	Difference
		(mil. dol.)	
Beef	$952.26	$1,078.13	$125.87
Pork	$560.21	$599.59	$39.38
Total			$165.25

Export Tax Policy

One of the largest impediments to international trade in feedgrains is the European Economic Community's (E.E.C.) Common Agricultural Policy (C.A.P.). A major provision of the C.A.P. is a variable levy which is the difference between a European support price (called the intervention price) for grain and the price of imported grain landed in Europe. Hence, the lower the price of imported grain, the higher the levy. Clearly, the impact of this policy is to reduce European demand for imported grain and to lower world (and U.S.) grain prices.

As one illustration of the interactions between the grain and livestock sectors, we asked: what if the U.S. imposed an export tax on feed grains? Specifically, our policy was formulated such that an export tax equal to the difference between the U.S. corn price and the internal E.E.C. price was applied on all feedgrain exported from the United States over our simulation period from 1969-1977. While this policy may have little effect on U.S. exports to the E.E.C. (since the U.S. export price would always be equal to the intervention price), it would have two effects on the U.S. grain sector. First, it would generate tax revenue for the U.S. treasury instead of the E.E.C. treasury; and second, it would reduce export demand for U.S. feedgrains in countries other than the E.E.C. Thus, more feedgrain at lower prices would be available to the U.S. livestock sector.

The results of this policy are presented graphically in

Figure 4. As expected, corn prices are generally lower than in the base run, except in 1974-75 when the landed price of U.S. corn in Europe was at or near the intervention price.

Average quarterly impacts on corn prices, feedgrain exports, total grain revenue and the export tax are presented in Table 3. The average impact on the first three variables is roughly the same in percentage terms as for the fixed exchange rate policy. Interestingly, however, the income generated by the export tax is greater than the loss in total revenue to producers.[7]

Table 3. Effects of Export Tax on U.S. Corn Prices, U.S. Feedgrain Exports and Total Revenue from Feedgrain Sales, 1969-1977 (Quarterly Average).

	Base Run	Export Tax	Difference
Corn Price ($/bu.)	$2.11	$1.97	$-.14
Feedgrain Exports (m.m.t.)	8.79	6.79	-2.00
Total Revenue (mil. dol.)	$3688.3	$3412.5	$-275.8
Export Tax (mil. dol.)		$344.9	

Livestock production would be stimulated by the export tax (Figure 4) and livestock prices would be lower because of lower feedgrain prices. It is again obvious that the impacts on livestock production increase throughout the simulation period and that the policy would result in relatively larger increases in beef production after 1978 because of the increasing beef herd in 1977 and 1978.

The impact of this policy on quarterly net revenue to the livestock sector is shown in Table 4. As can be seen, the gains to the livestock sector when added to the tax revenue generated by the policy far outweigh the loss in revenue to the feedgrain sector.

No Price Freeze and a Quota on Exports of Feedgrains

In April 1973, President Nixon imposed a freeze on retail prices of red meat and poultry. In August, he lifted the freeze for pork and poultry and announced that the freeze would be lifted for beef in October. His objective was to hold down inflation but his plan backfired and farm prices

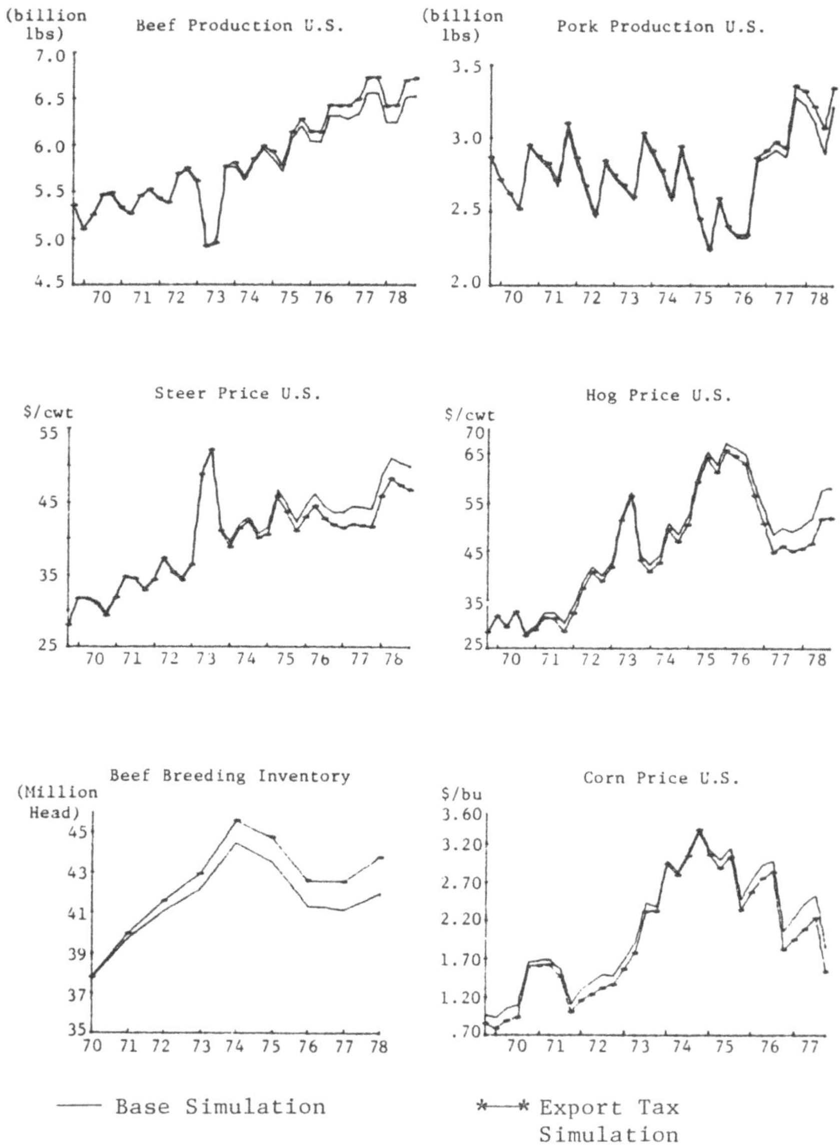

Figure 4. Export tax and base simulation, U.S. 1964-4 - 1978-4.

Table 4. Effects of Export Tax on Net Revenue to the Livestock Sector, U.S., 1969-1977 (Quarterly Average).

	Base Run	Export Tax	Difference
		(mil. dol.)	
Beef	$1030.97	$1138.90	$107.93
Pork	$ 545.98	$ 559.09	$ 13.11
Total			$121.04

rose. Beef producers began to hold back cattle during the late spring and summer in the hope that prices would increase more when the freeze was lifted. When the freeze was lifted on pork and poultry prices, they increased further (because cattle were being held off the market). Cattle producers held back determinedly. When the freeze on beef prices was lifted, cattle marketings returned to normal and all livestock prices fell.

During the period of the price freeze, the dramatic increases in livestock prices masked the (also) dramatic increases in feedgrain prices. The year was one of producer optimism in which record high steer and feeder calf prices provided a tremendous incentive to expand the breeding herd the following year. But the following year brought with it much lower steer prices, record high feedgrain prices and the beginning of four years of pain for beef producers.

This scenario led us to ask another "what if" question. What if the price freeze had not occurred and much stronger steps had been taken to avert increased feedgrain prices during 1974-75? To analyze this question, we did two things. First, the effect of the price freeze on the decisions by beef producers to hold cattle off the market is represented in the model by two "dummy variables" in the beef supply equation for the second and third quarters of 1973. These were relaxed in the analysis. Second, an export quota was placed on feedgrains such that no more than 15 percent of a given crop year's supply (production plus beginning stocks) could be exported from the U.S. The level of 15 percent was chosen because during the 22 years prior to 1973, an average of about 8 percent of each year's supply was exported with a maximum of 13.5 percent (1965). Hence, 15 percent would have, on an historical basis, allowed for a reasonable level of exports, but would protect the domestic livestock industry from large increases in grain prices.

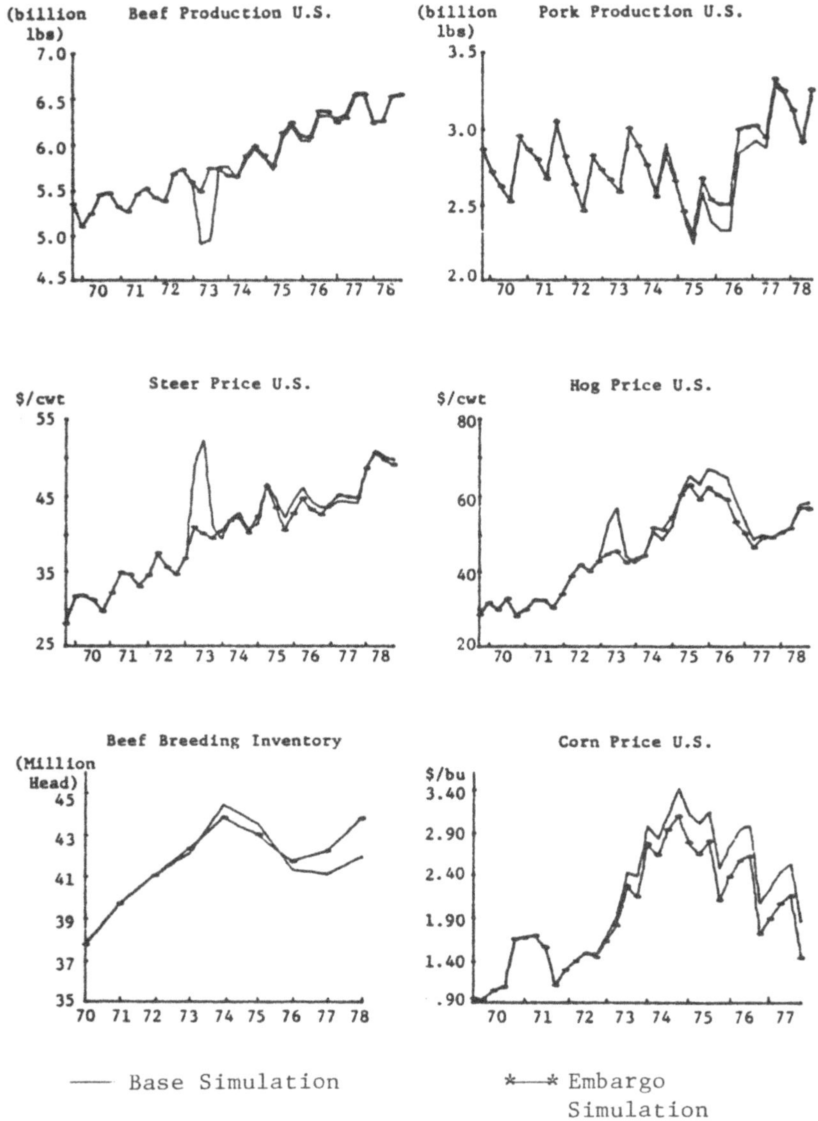

Figure 5. Embargo on feedgrains and base simulation, U.S. 1969-4 - 1978-4.

The expected effects of this set of policies are: (1) no price freeze and no hold back of slaughter cattle should increase beef supplies and lower beef and pork prices in late 1973; (2) the export quota should lower grain prices from 1972 onward.

Results of this analysis are presented graphically in Figure 5. As expected, domestic corn prices would have been lower than in the base run from 1972 onward. The impact of the policy on average quarterly U.S. corn prices, feedgrain exports and total revenue from sales of grain are continued in Table 5.

Table 5. Effects of No Price Freeze and Export Quota on U.S. Corn Prices, U.S. Feedgrain Exports and Total Revenue from Feedgrain Sales, 1971-1977 (Quarterly Average).

	Base Run	Export Quota	Difference
Corn Price ($/bu.)	$2.56	$2.32	$-0.24
Feedgrain Exports (m.m.t.)	11.62	7.64	-3.98
Total Revenue (mil. dol.)	$4123.2	$3969.59	$-153.6

The impacts of this policy on the livestock sector are interesting. Doing away with the price freeze would have increased beef supply in 1973 and caused prices of both hogs and cattle to be lower. As a result, cattle inventories would not have been built up as much in 1975. However, the lower (because of the embargo) feed prices in 1975, 1976, and 1977, would have caused a rapid expansion in the breeding herd in 1977 and 1978. While, after 1973, beef production would not have been much affected, it is again clear that beef production would increase after the simulation period because of the rapid build up in the breeding herd. Pork production would have increased somewhat in 1975, 1976 and 1977. As a result, hog and cattle prices would have been lower during this period.

The impacts of this policy on average net revenue to the livestock sector are presented in Table 6. The gain of $404.43 million to the livestock sector clearly exceeds the loss of $153.6 million in total revenue to the feedgrain sector.

Table 6. Effect of No Price Freeze and Export Quota on Net Revenue to the Livestock Sector, U.S., 1972-1977 (Quarterly Average).

	Base Run	Export Quota	Difference
		(mil. dol.)	
Beef	$807.12	$1138.60	$331.48
Pork	$572.27	$ 645.22	$ 72.95
Total			$404.43

A Buffer Stock Policy

While the foregoing policy scenarios are useful in illustrating the interrelationships between the feedgrain and livestock sectors, none is particularly realistic. Fixed exchange rates could not likely have been maintained in the face of the recent oil price experience. An export tax may be illegal in the United States and, the formulation used here would have provided the least impact when it was most needed--when grain prices were highest. The export quota would obviously have provided relief from high grain prices in 1974 and 1975, but it would also have made them even lower subsequently. A buffer stock policy has been suggested by many as an alternative to those discussed above.

Although there has been considerable research to determine the economic effects of national and international buffer stocks for wheat, the feedgrain market has largely been ignored. To the authors' knowledge only Taylor, et al.[8] have studied the effects of a buffer stock on the world feedgrain market. Their conclusions were: (a) The feedgrain market is intrinsically more stable than the wheat market and less affected by the operation of a buffer stock authority (B.S.A.); (b) Using a price trigger such that the B.S.A. would sell feedgrains at $3.30 per bushel and buy at $2.30 per bushel a buffer stock of 10 m.m.t. would reduce the standard deviation of price by 5 percent. This buffer stock was estimated to cost $368 million dollars over a 20 year period. The gains from trade for this buffer stock were estimated to be $426 million, of which $357 million would accrue to the United States; (c) A buffer stock larger than 10 m.m.t. increased the stability of prices only marginally; (d) Narrowing the range of the price triggers to $3.05 and $2.57 would decrease the variability of feedgrain prices by 12 percent but the buffer stock would run losses of $520 million.

Taylor, Sarris and Abbott conclude by saying that at most a rather small reserve scheme for feedgrains could be recommended and, given that most of the gains accrue to the feedgrain exporting countries, they may want to underwrite the costs of the B.S.A.

It should be noted that costs and benefits to the livestock sector were excluded from the analysis of Taylor, et al. While their analysis suggests that gains in stability to the feedgrain sector from a buffer stock policy may be small over time, it might well be enlightening to speculate about the impact of such a policy during the early 1970's. In 1972, during a period of low grain prices, the Nixon Administration made the decision to markedly reduce its Commodity Credit Corporation holdings of feedgrains in order to reduce treasury costs. As part of its program to carry out this decision, exports of wheat were subsidized. This provided an incentive for importing nations to import more of all grains. The effect of this policy was to put pressure on an already low priced market in 1972, encourage exports, increase domestic feedgrain prices in 1973 and make available less stocks in 1974 when export demand, coupled with a smaller domestic crop drove domestic prices 230 percent higher than in 1971-72. As has been shown, these grain prices had substantial impacts on the livestock sector, subsequently. Had a buffer stock policy formulated along the lines suggested by Taylor et al. been followed during this period, grain prices would likely not have risen to the heights they did, and the subsequent adverse consequences to the livestock sectors would likely not have been so drastic. Hence, it can be hypothesized that if the impacts on the livestock sector of a feedgrain buffer stock were included in the analysis, the benefit/cost ratio would be positive.

Conclusions

Despite the several limitations noted above in both the modeling process we used and in analyzing the impacts of policies, two tentative conclusions can be offered. First, the analysis suggests that changes in feedgrain policy can have impacts on the livestock sector which are as great or greater than their impact on the feedgrain sector. This is demonstrated well by the no price freeze and export quota policy. Even given the limitations of the rudimentary manner in which the impacts of that policy were measured, the economic gains to U.S. society from a policy which lowers and stabilizes feedgrain prices are understated by our analysis. No account, for example, was taken of the impacts of these policies on the poultry industry, on the welfare of consumers and on the multiplier effect of a larger and more stable livestock industry to the U.S. economy.

In a similar vein, the analysis of an export quota and our discussions of a buffer stock policy suggest that a stable feedgrain sector is an important, if not the most important, factor required to stabilize the livestock sector.

These tentative inferences, if they can be confirmed by more thorough analyses, are of obvious importance to informed policy decisions.

The second conclusion lies in the extreme time lags between a policy decision which affects the feedgrain market and its consequent impacts on the livestock--particularly the beef--sector. In each of our policy scenarios, the first impacts on the feedgrain sector would have occurred during the early 1970's, but the major impacts on beef supply would not have occurred until after the simulation period ended in 1978. These time lags are important to the policy making process. If they are not recognized or are discounted for political reasons, the person responsible for the original policy decision may not be accountable for its long term outcome. For example, the decisions made by the Nixon Administration during the early 1970's to reduce grain inventories had the immediate effect of raising grain prices. But the longer term impacts of higher grain prices on the beef sector were not felt until 1975-1977 when the beef herd was substantially liquidated and from 1978 onward as the impact of that liquidation becomes obvious in the consumer price index. These events have and will occur during the Ford and Carter administrations. Similarly, while it has nothing to do with grain policy, Mr. Carter's decision to increase beef import quotas during 1978, just at the time when many in the beef industry were regaining confidence and beginning to rebuild the beef herd, may not have its major impacts until after the election of 1980.

References and Notes

1. Anthony S. Rojko, "The Economics of Food Reserve Systems," American Journal of Agricultural Economics, 57, No. 5, pp. 866-872, Dec. 1975.

   Lance Taylor, A. H. Sarris and P. C. Abbott, Grain Reserves, Emergency Relief, and Food Aid, Washington: Overseas Development Council, prepublication version, 1977.

   Shlomo Reutlinger, "A Simulation Model for Evaluating Worldwide Buffer Stocks of Wheat", American Journal of Agricultural Economics, 58, No. 1, pp. 1-12, Feb. 1976.

   Anthony C. Zwart, An Empirical Analysis of Alternative Stabilization Policies for the World Wheat Sector, unpublished Ph.D. thesis, University of Guelph, 1977.

2. The cause of this problem has subsequently been found to be errors in two of the data series used in the analysis.

3. For an overview of the approach used in modeling the livestock sector, see: Larry Martin and A. C. Zwart, "A Spatial and Temporal Model of the North American Pork Sector for the Evaluation of Policy Alternatives," American Journal of Agricultural Economics, 57, No. 1, pp. 55-66, Feb. 1975; and, Larry Martin and Karl D. Meilke,"Commodity Modeling for Policy Analysis in a Canadian Context," Agricultural and Food Price and Income Policy for the U.S. and Implications for Research, ed. Robert Spitze, Special Pub. 43, University of Illinois, Aug. 1976, pp. 107-117.

4. Harry deGorter of Agriculture Canada provided all of the data used in the feedgrain model and an initial specification of the demand side of the model. The specification of the feedgrain model is similar to that reported in Karl D. Meilke, "An Aggregate U.S. Feed Grain Model," Agricultural Economics Research, 27, No. 1, pp. 9-18, Jan. 1975.

5. In simulating the policy options, we iterated between the two models to capture at least some of the simultaneity in the two sectors.

6. The discussion in this section and the supply block for the livestock model is based on Larry Martin and Richard Haack, "Quarterly Beef Supply Response in North America," Canadian Journal of Agricultural Economics, 25, No. 3, No. 1977; and Larry Martin and A. C. Zwart, op. cit.

7. The export tax policy undoubtedly overstates the level of U.S. exports and tax revenue since it was assumed that feedgrain supplies in the rest of the world would not increase in response to the higher U.S. export price for feedgrains.

8. Lance Taylor, A. H. Sarris and P. C. Abbott, Grain Reserves, Emergency Relief, and Food Aid, Washington: Overseas Development Council, prepublication version, 1977.

Nathan E. Smith

3. Opportunities for Forage, Waste and By-Product Conversion to Human Food by Ruminants

Introduction

Availability of food (quantity and quality) for the human population is and will be a major world concern. Because of the constraint of available land mass, resolution of optimum use of land resources and crops therefrom is required for maximum productivity of human food. Of the total world land mass, it is estimated that about 11.5% is arable and suitable for cultivated crop production while more than 50% is either grassland or forest (Table 1). It is quite apparent that utilization of both types of food producing lands is essential to maximal human food production. Much attention has been focused on the use of harvested crops from arable land as feeds for livestock. Major emphasis has been placed on apparent inefficiencies involved in this scheme of human food production. Some of this attention is appropriate. However, consideration of a multitude of additional factors is required for critical analysis and planning for future food supplies.

Table 1. World Land Resources*

Type	Hectares	% of Land Mass
Arable	1.50×10^9	11.5
Grassland	3.05×10^9	23.3
Forest and Woodland	4.16×10^9	31.8
Other	4.37×10^9	33.4
Total	13.08×10^9	100.0

*From FAO Production Yearbook, 1976.

A large portion of tillable land resources are either best suited to forage production or require forage cropping in rotational schemes for maximum crop output. In addition, much of this resource is not well suited (because of land-type, climate, etc.) to production of crops providing highest quality human foods such as wheat, rice, soybeans, etc., but rather to crops somewhat less digestible, of poorer quality or less palatable to humans such as corn, barley, oats, sorghum, etc. The production, harvesting and processing of grain, vegetable, fruit and nut crops from arable land results in the availability of considerable quantities of crop residues and by-product feedstuffs not directly edible by humans.

Ruminants, because of extensive microbial digestion, are unique in their ability to convert poor quality feedstuffs into high quality human foods (meat and milk). Pasture and forage crop-lands have historically been the basis of ruminant production systems. It is quite obvious that if we are to produce human food from these vast resources it will necessarily come about through harvest and utilization by ruminant animals. The potential for utilization of other resources by ruminants in producing human food is not quite so obvious.

Feed Resources and Use in the U.S.

Land resources and utilization for crop and food production in the United States differs considerabley from the total world situation. However, since relatively complete data are available, examination of food and feed resources and utilization in the U.S. serves as a basis for gaining greater insights and knowledge relative to the potential for ruminants in world food production systems.

Of the total land mass in the U.S. (including all 50 states), about 15% is utilized for harvested crop production and 40% is pasture land (Table 2). Both of these values are slightly above world average and emphasize the point that vast quantities of poor quality feedstuffs (pastured crops) are available for conversion to human food by ruminant animals. Basic questions arise relative to optimum utilization of the 135 million hectares of harvested cropland.

Grain and Oilseed Production and Use

Twenty percent of harvested cropland was planted to oilseed crops in 1976 (2) with soybeans accounting for over 80% of this total. Fifty five percent of this cropland was used for the production of grain crops in 1976 with wheat and corn as the major crops (Table 3). About 60 percent of corn, barley, oats and sorghum produced was utilized directly for animal feed. The relatively high proportion used for animal feeds undoubtedly reflects the poor quality and/or lower acceptability of these grain crops for direct human consumption.

Table 2. Food or Feed Producing Lands (U.S.-1969)*

Type	Hectares	% of Total Land
Cropland (harvested)	135×10^6	15
Pasture:		
Cropland	36×10^6	
Permanent or Range	300×10^6	
Forest	25×10^6	
Total pasture	361×10^6	40

*From U.S.D.A. Agricultural Statistics, 1976. Includes all 50 states.

Table 3. Cereal Grain Production and Utilization (U.S.-1976)[a]*

		Metric Tons	
Crop	Hectares	Produced	Used directly for Livestock and Poultry Feed
Corn	27.1×10^6	78.9×10^6	47.3×10^6
Wheat	28.4×10^6	29.2×10^6	1.1×10^6
Barley	3.6×10^6	4.1×10^6	3.0×10^6
Oats	5.7×10^6	4.1×10^6	3.0×10^6
Rye	0.4×10^6	0.2×10^6	0.1×10^6
Rice	1.2×10^6	1.2×10^6	—
Sorghum	6.1×10^6	9.2×10^6	6.4×10^6
Total	72.5×10^6	126.9×10^6	60.9×10^6

*From U.S.D.A. Agricultural Statistics, 1976.

Economic factors also exert considerable influence on the utilization of these, or any other feed or food crops. Ruminant production industries are the recipients of much criticism because of this direct use of grain crops and the apparent competition with man's food supply. Some of this criticism may be appropriate. However, it must be remembered that ruminant production systems are based on economic principles. Feedstuffs will be utilized for these animals only if they are cheaper sources of nutrients than other feeds and will result in greater dollar margins. If grain crops are of higher quality and more acceptable to humans, the cost is higher and very little is used directly for animal feed. This is readily apparent, for example, in the case of wheat and rice where less than 4 percent of the grain crop is used directly for livestock and poultry feed (Table 3). Even if grain and oilseed crops are used directly for human food, significant portions of the harvested crop do not appear in the final product and thus become an available resource for other routes of utilization for food production.

By-Product Feeds

Processing of grains and oilseeds for human foods results in availability of a wide array and considerable quantities of by-product feedstuffs for utilization as animal feeds. In 1976, over 16 million metric tons of such by-products were utilized as feeds for livestock and poultry in the United States (Table 4). This was over 20 percent of total feed concentrates fed to these animals. By-product feeds are the result of separation or extraction of higher quality portions of fruits, vegetables or seeds for human consumption and thus are usually of relatively poor quality or low acceptability for human foods. As feeds for ruminants, however, they are often equal to, or even exceed, the nutritional value of the original harvested crop. Energy available from these by-product feed resources in the U.S. is approximately 50 billion Mcal metabolizable energy (ME) if fed to ruminants.

Whole Crops vs. Grain

An additional consideration to be made when evaluating production of grain crops, and ultimate use by man directly or through animal production is that considerable high quality ruminant feed is produced in addition to the seed, especially if the crop is harvested prior to maximum maturity. Let us examine the corn crop, for example. If the crop is allowed to mature and the grain crop harvested, the yield may be in the order of 7,500 kg grain dry matter/hectare or about 26,000 Mcal ME/hectare. However, the whole plant may be harvested slightly prior to maturity (when the stalks, leaves, etc. have excellent feed value), chopped and stored and fed as silage. The resulting corn silage is an excellent feed for ruminants and the expected yield would be about 16,000 kg dry matter/hectare or about 42,000 Mcal ME/hectare.

If one assumes a liberal 60% direct utilization by humans of corn grain, human food energy from the grain crop is equivalent to 15,600 Mcal ME/hectare. On the other hand, harvesting the whole crop as corn silage and feeding the silage to lactating cows would result in approximately 30,000 kg milk or 22,500 Mcal ME as human food (an additional 45% output of human food energy). Similarly, other grain crops can be harvested as whole plant silage providing high quality ruminant feeds. It would appear that critical analysis and planning for food production systems in the future should include consideration of alternatives such as whole crop vs. grain crop harvest.

Concentrate Feeding and Productive Efficiency

An important factor in consideration of optimum utilization of feed grains and by-products, which is too

Table 4. Vegetable By-Product Feeds Fed to Livestock and Poultry (U.S. - 1976)*

Feed	Metric Tons
Soybean meal	7.08×10^6
Cottonseed meal	0.59×10^6
Peanut meal	0.14×10^6
Linseed meal	0.05×10^6
Wheat mill feeds	2.13×10^6
Gluten feed and meal	0.68×10^6
Rice mill feeds	0.23×10^6
Brewers grains (dry)	0.14×10^6
Distillers grains (dry)	0.18×10^6
Beet pulp (dry)	0.77×10^6
Other	4.21×10^6
Total	16.20×10^6

*From U.S.D.A. Agricultural Statistics, 1976.

often overlooked, is the potential increase in efficiency of human food production realized by supplementing forage or pasture with these feed concentrates in feeding programs for ruminants. The effects of combining feed concentrates with forages on efficiencies of meat and milk production from ruminants have been discussed in detail by several authors (3, 4, 5, 6). Dramatic increases in efficiency of production are realized with optimum combinations of feed concentrates and forages and, in fact, human edible food output may be greater than human edible food input in such feeding systems (see chapter 1). Reasons for these effects include the relatively large portion of feed energy and protein being utilized for animal maintenance in forage feeding programs because the bulk of these feeds limits consumption. This results in low productivity and efficiency. Addition of high energy (concentrate) feeds increases energy and protein intake and alters digestion and metabolism leading to greater animal efficiency. It is obvious that such factors must be considered in planning for optimal utilization of crops produced on arable land for net human food production.

Crop Residues

In addition to human edible and by-product foods and feeds harvested and available from grain, oilseed, vegetable, fruit and nut crops, substantial quantities of crop residues are also produced. In the harvesting of grain crops, for example, large quanitities of straw, stover and other residues are available. Similarly, only portions of vegetable and root crops are harvested and processed or used directly for human consumption leaving considerable wastes or residues. For most of these crops, the quantity of residue is at least equal to the quantity of material harvested (7, 8). Such residues represent a tremendous potential feed source (especially for ruminants), but, in general, have not been well utilized in food production systems for several reasons.

Physical and economic factors and problems have been major deterrents to maximum utilization of these feeds. Handling, storage and feeding of crop residues is often difficult because of the nature of methods used to harvest the primary crop. Harvesting methods for many crops leave residues scattered and in close contact with the soil presenting problems in removal and adequate field drying for storage. However, significant advancements in methods for residue harvest and handling have been made in recent years. Currently, it appears that the primary limiting factor in residue utilization is economics (the value of residues relative to other available feedstuffs). This is particularly true with grain crop residues.

The plant, at the time of grain harvest, is mature and thus residues from these crops are characterized by low nitrogen, high cell wall content, cell wall lignification, and, in some residues, high silica content. In general, such residues represent relatively poor quality feedstuffs even for ruminants. Feeding of grain crop residues to ruminants or other animals requires considerable dietary nitrogen supplementation compared to feeding programs based on traditional forages. Digestibility of these residues is somewhat variable, but is very low in comparison with forages such as grasses and legumes which are harvested at less mature stages of growth. Energy values for residues such as corn cobs and wheat or oat straw fed to ruminants are about 1.6 Mcal ME/kg dry matter compared to 2.3 to 2.5 Mcal ME/kg dry matter for good quality grass or legume forage (9). Such differences take on particular importance in feeding programs for ruminants where significant productivity (meat or milk production) is a required output. Diets containing large quantities of such low energy feedstuffs can be used relatively satisfactorily for maintaining mature ruminants but better quality feedstuffs and greater dietary energy concentrations are required for productivity. This imposes a limit on extent of residue feeding unless methods of improving quality are adapted.

Residue Processing

Considerable research emphasis in recent years has been placed on development and evaluation of techniques to enhance the digestibility of grain crop residues and/or improve performance of ruminants fed such feedstuffs. The work has been reviewed in depth (8, 10). The primary processing methods studied have been physical or chemical treatments, or combinations of both.

Physical processing techniques have included various methods of chopping, grinding, pelleting or cubing. Most studies with straw residues have shown considerable increases in feed efficiency and ruminant productivity in response to such treatments (10). The response appears to be due primarily to increased intakes of processed straw residues resulting from alterations in physical form and increased density of these feedstuffs.

Several chemical treatments of grain crop residues have been evaluated for potential enhancement of digestibility

(8, 11, 12). Chemicals used in ruminant feeding studies have included sodium hydroxide, ammonium hydroxide, calcium hydroxide, and potassium hydroxide. The majority of studies have been conducted with sodium hydroxide, and, in general, it appears to be the most effective in increasing digestibility. However, in several studies where chemical treatment was combined with physical processing such as ensiling or pressure treatment, ammonium hydroxide appeared to be as effective as sodium hydroxide (8, 13, 14). In addition, the nitrogen from the added ammonium hydroxide appears to be efficiently utilized by ruminants (14). This could be an important factor in future considerations of chemical treatment of straw residues since treatment with ammonium hydroxide would help supply required dietary nitrogen supplementation as well as increase digestibility and utilization of these feedstuffs.

Digestibility and feed efficiency of most grain crop residues can be increased by at least 20 to 30 percent with proper chemical and physical treatments (8, 10). Consideration of this, in light of the magnitude of available residues, reveals the tremendous potential for food production if these residues were maximally utilized in ruminant feeding programs. Approximately 200 million metric tons (dry weight) of grain crop residues are produced in the United States each year (2). If this resource was maximally (100%) utilized for ruminant feeding and the digestibility or energy value for ruminants was increased by 20% (to 1.9 Mcal ME/kg dry matter), total energy available from residues would be in the order of 380 billion Mcal ME. Adding the 50 billion Mcal ME from by-product feeds yields a total potential of 430 billion Mcal ME available each year in the U.S. from these two resources. Analyses (15, 16) indicate that total feed energy requirements for beef or dairy production systems are about 20 Mcal ME/kg gain and 3.5 Mcal ME/kg milk produced. Thus, the by-product feeds and grain crop residues in the U.S. have the potential for production of 21 billion kg of beef gain (about 60 kg carcass beef per person) or about 120 billion kg milk (approximately twice the total produced in the U.S. each year). It is obvious that, in the United States, crop residues are not being well utilized in present food production systems.

Extrapolation to the World Situation

Extrapolation of these considerations and calculations to the world situation staggers the mind. Total world production of grain crops (and thus of these residues) is 10 to 15 times that of the United States (1, 2). Residues from other harvested crops represent additional significant quantities of potential feedstuffs. World land resources for food production are not increasing, nor can they be expected to change drastically in the future. At present, there is about 0.35 hectare arable land per capita in the world. Based on projections for world population increases, this will decrease by some 30 percent (to less than 0.25 hectares per capita) by the year 2000. Expectation of 100 percent utilization of all food producing resources (especially crop residues) is probably unreasonable. However, it is also obvious that considerable pressure will be put upon food production systems in the future with the basic requirement being that of more fully utilizing total food production resources.

Summary

Coping with world food problems in the future will not be an easy task. With limited food production resources, critical analyses and planning will be required. We have, at our disposal, vast areas of grass and forest lands for use in food production systems. Production from arable land, in addition to crops or products directly consumed by humans, includes by-product feeds and large quantities of crop residues and wastes in addition to alternative uses such as whole plant harvest as silage. Emphasis has been placed upon production of human edible food from arable land resources, and rightly so. However, it would be absurd not to recognize the potential and importance of these other available feedstuffs. Ruminants, because of their ability to convert these resources to high quality human food, can and do play an extremely important role in the human food production chain. It is essential that all factors and alternatives be considered in planning for the future. This includes ruminants and the vast array of their present and potential feedstuffs.

References

1. FAO. 1977. Production Yearbook, 1976 Rome, Italy: FAO.

2. USDA. 1976. Agricultural Statistics. U.S. Department of Agriculture.

3. Baldwin, R. L., B. D. Slenning and M. Ronning. 1978. A visualization of the livestock industry in the world perspective. In Nutrition in Transition, Proc. Western Hemisphere Nutrition Congress V, ed. by P. L. White and N. Selvey. American Medical Association, Monroe, Wisc.

4. Reid, J. T. 1974. Energy metabolism of the whole animal. In The Control of Metabolism, ed. by J. D. Sink. Pennsylvania State Univ. Press, University Park.

5. Blaxter, K. L. 1962. The Energy Metabolism of Ruminants. C. C. Thomas, Springfield, Ill.

6. Smith, N. E. 1978. The efficiency of energy utilization for milk production from intensive and extensive systems. Proc. 3rd. World Congress on Animal Feeding, Madrid. p. 123.

7. Knutsen, J. 1976. Crop Residues in California. Univ. of Cal. Leaflet 2872.

8. Klopfenstein, T. 1978. Chemical treatment of crop residues. J. An. Sci. 46:841.

9. National Research Council. 1978. Nutrient Requirements of Dairy Cattle, 5th Edition. National Academy of Sciences, Washington, DC.

10. Anderson, D. C. 1978. Use of cereal residues in beef cattle production systems. J. An. Sci. 46:849.

11. Chandra, S. and M. G. Jackson. 1971. A study of various chemical treatments to remove lignin from coarse roughages and increase their digestibility. J. Agric. Sci. 77:11.

12. Garrett, W. N., H. G. Walker, G. O. Kohler and M. R. Hart. 1976. Feedlot response of beef steers to diets containing NaOH or NH_3 treated rice straw.

proc. 15th Calif. Feeders Day, Univ. of Calif. at Davis. p. 39.

13. Penn, T. P. 1978. The effect of feeding sodium hydroxide and ammonium hydroxide treated rice straw on rumen fermentation in the bovine. M.S. Thesis, Univ. of Calif. Davis.

14. Collar, L. S. 1978. Effect of ammonium hydroxide treatment of rice straw on digestibility and NPN utilization by ruminants. M.S. Thesis. Univ. of Calif., Davis.

15. Fitzhugh, H. A. 1978. Bioeconomic analysis of ruminant production systems. J. An. Sci. 46:797.

16. Reid, J. T. 1970. Will meat, milk and egg production be possible in the future? Proc. Cornell Nutrition Conference, Buffalo, NY, p. 15.

Henry S. Bayley

4. Opportunities for Waste and By-Product Conversion to Human Food by Non-Ruminants

Relationship of Food, Animals and Man

Any consideration of the relative importance of ruminant and domestic animals in meat production in relation to feed consumed by the animals must be prefaced by consideration of the biological significance of ruminant versus non-ruminant digestion. The symbiotic relationship between the animal and the microflora which inhabit the proximal region of the digestive tract in ruminants represents a very sophisticated system for food utilization. This system allows ruminants to use cellulose as an energy source whereas non-ruminants are either unable to use it or are only able to use cellulose to a limited extent. Since cellulose is the most abundant organic compound in the biosphere, this difference between ruminants and non-ruminants is central to any consideration of their food supply.

Non-ruminant animals have adapted to their environment in a number of ways. In general, herbivorous, simple stomached animals are small in size and are restricted in their choice of habitat. Carniverous animals which prey upon ruminant animals are able to obtain energy from cellulose indirectly, through the ruminants. Man has developed a special relationship with ruminant animals, optimizing the conditions for their productivity by husbandry, and in developed economies by providing them with forage crops as feed. This association between man and domesticated ruminants has allowed the exploitation of cellulosic material as a food source for man. There has been an association between man and ruminant animals for a considerable part of human history; the animals providing a means of harvesting land for which man did not have the technological resources to crop for direct human food production. It is only in very recent time that there has been the opportunity to produce food crops in such abundance that the parts of the plants which are suitable for direct consumption by man could be considered animal feedstuffs. It could be concluded that man's first association with animals as a means of obtaining food from wastes, began by using animals to harvest 'waste' lands, and only in very recent times have human foods such as cereal grain been diverted to animal feeding.

Although the digestive capabilities of simple stomached animals prevent the use of the cellulosic parts of plants, many parts of plants contain enough starch, oil or protein to make them valuable as foods for man making it even possible for human societies to exist using plants as the basis for their food. Only a limited part of the plant is consumed, usually the fruit, seed, or vegetative storage organ: the remaining portion of the plant being discarded. In cultures which use both plants and animals as food, some parts of the animal carcass may also be rejected for consumption and discarded. The completeness with which foods are consumed depends upon the customs of a particular society, and in general, bears an inverse relationship to the prosperity of the group. It is not surprising that most of what is known about the development of civilization has been learned from examining the waste dumps which are invariably associated with human settlement. In modern, industrial societies there is always some part of the food material which is rejected for direct consumption by man and this constitutes the 'waste' which may be an important feed resource.

Foods, Feedstuffs and Nutrition

It is the task of the nutritionist to provide a quantitative assessment of the nutritional values of wastes. This is the first stage in evaluating the economic feasibility of employing such materials in feeding systems for domestic animals. Such an evaluation requires comparison of the nutrient contents of the wastes with the nutrient requirements of each species of farm livestock, an additional aspect of such an evaluation involves consideration of the physical and chemical forms of the waste in relation to the prehensive and digestive abilities of various animals.

The need to develop a quantitative system for classifying and evaluating feeds has long been recognized by animal nutritionists. Henneberg and Stohman, working in the early part of the nineteenth century, introduced the 'Weende' system of feed analysis which describes feeds in terms of their proximate components: moisture, ash, fat, fibre, crude protein and nitrogen free extract. This system of describing feeds is still important in spite of the fact that none of the proximate components describes a well defined chemical entity.

The utility of the Weende system is due to the quantitation of the crude fibre part of the feed; this approximates to the cellulose which cannot be digested by simple stomached animals. The nitrogen free extract gives an estimate of the amount of carbohydrate which can be digested by simple stomached animals. Thus on the basis of the Weende analysis, feeds can be directed towards ruminants or non-ruminant animals to obtain maximum productivity from a given feed resource. In more recent times feed classification has been formalized by the National Academy of Sciences, on the basis of the Weende feed analysis system as follows:

dry roughage: feeds with more than 18% crude fibre (on a dry basis)
protein supplements: feeds with more than 20% crude protein (on a dry basis)
energy feeds: containing less than 18% crude fibre and less than 20% crude protein (on a dry basis).

Thus any waste should be subjected to a proximate analysis in order to indicate its potential value as a feed.

Now that animals' requirements in terms of specific nutrients are reasonably well established it is necessary to consider the levels and availabilities of specific nutrients in wastes being evaluated as potential livestock feeds. There are over 40 elements and compounds which are needed by animals to maintain their health, although some of these compounds can be obtained endogenously from within the digestive tract where they are produced by the microorganisms. For convenience, the nutrients are considered under five headings: minerals, vitamins, fat, protein and energy.

Minerals are needed by the animal for bone function, as enzyme activators, for the transport of oxygen and for the transfer of information in nervous tissue. Quantitative needs for different minerals vary widely. Some minerals including calcium and phosphorus are required in fairly large quantities. The needs for some other minerals are so small that it is difficult to produce a diet which is deficient in these minerals because of their ubiquitous nature. However, not all minerals present in the diet can be absorbed from the digestive tract because of the formation of insoluble complexes with other components of the diet, for example zinc availability is depressed by high levels of calcium in the diet. A characteristic dermatitis, parakeratosis, occurs in pigs receiving diets with excessive levels of calcium. Another problem regarding minerals which must be considered is that there is often a narrow margin between required levels

Table 1. Nutrient Contribution of Corn, Soybean Meal and Mineral Supplement to a Typical Pig Feed.

Feed Ingredient	Nutrient Levels				
	Level in Diet (g/kg)	Digestible Energy (kcal/kg)	Protein (g/kg)	Calcium (g/kg)	Phosphorus (g/kg)
Corn	820	2877	84	0.16	0.82
Soybean meal	150	495	66	0.48	0.35
Supplement	30	-	-	4.36	2.83
Totals	1000	3372	150	5.00	4.00

and levels which are toxic, more than 250 mg of fluoride per kg of diet has adverse effects on swine. These observations mean that an evaluation of a 'waste' as a feedstuff must consider not only levels of the nutritionally important minerals but also possible effects of known interactions among them and their nutritional or toxic effects.

Many vitamins are unstable, being destroyed by exposure to heat, light, oxygen or moisture, thus there are few 'wastes' which can be considered as dependable sources of these nutrients. Exceptions are those materials which have suffered minimal destruction of cell and tissue structure. Thus, cereal milling by-products retain their water soluble vitamins, whereas the oilseed meals which have been heated and often solvent extracted have lost much of their vitamin content.

Animals synthesize fatty acids extensively but the linoleic acid 'family' of fatty acids must be provided by the diet. Thus it is important to determine levels of these essential fatty acids in wastes before they are used in feeding systems. In practice the expensive assays involved in measuring vitamin contents of wastes, and the instabilities of vitamins, and unsaturated essential fatty acids coupled with high variability in wastes leads to the observation that many wastes are undefined as sources of these nutrients. As a result, unfortunately, contributions which they could make to the diet with respect to these nutrients are often discounted.

Crude protein is merely a measure of the amount of nitrogen in organic combination, and as such gives no indication of the array of amino acids which constitute the protein and which are required by the animal. Also, protein structures in waste materials may resist digestion. This means that elaborate experiments to measure the availability and nutritional value of proteins in wastes must be carried out to fully characterize this component.

The useful energy value of feed is reduced by crude fibre, particularly for simple stomached animals, but is increased by fat. This makes the energy value of a waste the most easily predicted of the nutritional parameters. Much of the economic incentive to incorporate wastes into animal feeding systems arises from their value as energy sources.

In well developed economies, meat production from pigs and poultry is based upon the use of grains supplemented with protein, vitamins and minerals. Table 1 shows that a typical pig feed contains over 80% corn with soybean meal used as a protein supplement. In this example the corn supplies almost all the digestible energy of the diet and over half of the protein. Corn makes very minor contributions of calcium and phosphorus which are needed in a balanced pig diet. The soybean meal supplies some energy to the diet, but its main function is to supply protein. The supplement is needed to provide the major part of the calcium and phosphorus. This simple calculation for a three component diet considers only four of the forty nutrients, but the same principle applies to the assembly of balanced diets from feedstuffs of known composition to fulfill the nutrient needs of a particular species of animal. The arithmetic of such dietary assembly is usually delegated to a computer. An automated system[1] has been described for such calculations using 40 nutrients and any number of feed ingredients which leaves the nutritionist free to make decisions based upon nutrient balance, rather than upon cost considerations.

Using this type of diet under normal husbandry conditions, it would need about 300 kg of feed to produce one pig of 100 kg liveweight; this corresponds to the following inputs:

Digestible energy	1,000,000 kilocalories
Crude protein	45 kg
Calcium	1.5 kg
Phosphorus	1.2 kg

It is against these nutrient inputs, and their cost that the economics of waste utilization must be weighed.

Wastes Well Established as Feed Resources

Wastes arising from crops are either dry and relatively stable residues from grain production, such as straw, wheat offals and rice polishings, or high moisture residues from fruit and vegetable processing such as citrus pulp. They are predominantly cellulosic which is usually why they have been rejected for direct human consumption. However, they often contain sufficient amounts of protein, starch, pentosans or fat to make them potentially useful feedstuffs for simple stomached animals. Some of these by-products are well established as animal feeds. As an example, wheat flour production extracts 70% of the grain as flour leaving 30% as milling residues. The economic value of this residue is maximized by fractionating it into wheat bran, wheat shorts and wheat middlings. These are characterized by their levels of crude fibre and protein: middlings containing 18% protein and only 2% crude fibre is a useful source of protein and energy for swine and poultry whereas bran containing 10% crude fibre is better utilized in ruminant feeding.

Citrus pulp has a high crude fibre level (18% of the dry feed) and it is customary to utilize this product in ruminant feeding. However, it does contain some protein (6%) and recent evaluations have shown that swine can utilize citrus pulp as a source of both protein and energy; the energy being derived from the digestion of hemicellulose and pentosans. Substitution of citrus pulp for cereal in pig diets does result in some decline in growth rate[2]. Decisions concerning the use of such fibrous crop residues in pig diets, even if it means a lowering of growth rate, depends upon the relative costs of wastes and cereals.

Vegetable oil production is important, and in temperate zones is based upon soybeans, cottonseeds and rapeseeds. Oil is extracted from the seeds in central processing factories by exposing the seeds to a combination of pressure, heat, and chemical solvents. This leaves the remainder of the seed as residues which contain substantial amounts of protein. The large quantities of soybeans processed means that soybean meal is one of the predominant sources of supplementary protein used in animal diets. The development of technology necessary to produce soybean meal as a consistent source of quality protein has been an example of co-operation between nutritionists and food processing engineers. Similar efforts are being exerted to develop rapeseed meal as a consistent source of protein for animal feeding. In this case, plant breeders are also adding their expertise to change the chemical characteristics of rapeseeds to enhance the nutritional values of both rapeseed oil and rapeseed meal. The oilseed meal residues play such an important part in animal feeding systems that it would be impossible to sustain the levels of meat production enjoyed by developed societies without such a source of feed protein. However, the monetary value of the vegetable oils exceeds the monetary value of the residue, so that in spite of their importance, the oilseed meals are properly considered as by-products.

Wastes arising from the preparation of animal products for consumption have a high moisture content and are unstable. Such products must be either used as feedstuffs in their fresh state, or must be processed to enhance their stability. Slaughtering of animals for meat is associated with the rejection of parts termed inedible. These are dehydrated, defatted and ground, to produce meat meal, meat and bone meal, blood meal and tallow. There is a well developed technology to insure the production of stable, sanitary products by the rendering industry. Fishery wastes are particularly unstable being subject to both autolysis and microbial spoilage. These by-products are usually available only during particular seasons of the year, making it economically unattractive to commit large investments to rendering facilities. Some efforts are being expended to assess the process of ensiling such wastes at low pH as an economical means of preservation. Animal by-products (meat and fish products) are particularly valuable as protein sources, not only because of their high protein content, but because the amino acid composition of the protein is well suited to the needs of most animals for growth.

Cheese manufacturing results in the production of large amounts of whey which is customarily used in pig feeding; however, whey contains only 5 to 8% dry matter making it expensive to transport or to dehydrate. The use of whey in pig feeding is usually associated with diarrhea, so that although whey contains a valuable protein which constitutes 12% of the dry matter, its usefulness as a feed resource is limited because of the pig's limited ability to digest the lactose which makes up the major part of the dry matter (75%). Recognition of the cause of the diarrhea[3] may lead to the development of feeding systems which facilitate the digestion of lactose and enhance the usefulness of whey as a feedstuff.

The beverage industry is a source of wastes which have traditionally been used as feeds; brewers' grains, brewers' yeast and distillers' solubles. Wastes which arise from the application of modern technology to enhance consumer convenience such as spent coffee grounds should also be included here. The latter remain in large quantities from the production of instant coffee and are useful feedstuffs because of their fat content. The alcoholic beverage by-products have been subjected to microbial fermentation and contain substantial amounts of vitamins. In addition, it is often claimed that such products contribute 'unidentified growth factors' to the animals' diet. Unfortunately, by-products of the dairy and beverage industries contain large amounts of water and must either be fed in the fresh (or undehydrated) state which limits the distance they can be transported, or the considerable costs of drying must be incurred. Such products have a place in animal feeding systems. Economic barriers to their utilization mean that it is often most economical to dump them into the nearest water course; such practices are no longer socially acceptable so efforts are being expended to recover their feed value in the most economic manner.

The wastes considered to this point are all being utilized to some degree in animal feeding. Technology exists to incorporate each product into some system of animal production. The decision to dump the waste or use it as feed is largely dictated by economic considerations. Environmental constraints on dumping are leading to increased efforts in the development of economical methods for utilization of these products in animal feeds. Efforts to further increase meat production using existing feed resources calls for the identification of wastes which are either not being used as feed, or wastes which can be utilized with greater efficiency by the application of new technology. One of the most obvious sources of nutrients is domestic kitchen waste. Currently the disposal of such material in urban centres imposes a major burden upon sewage treatment facilities and general refuse disposal systems. It should be emphasized that the earliest function of the pig in agrarian economies was to utilize just this type of waste. Unfortunately the high water content, the rapidity of microbial spoilage and the organizational difficulties of collecting domestic garbage make it economically unattractive as a source of nutrients. There are well established pig production facilities which use garbage as the major feed input, usually based upon garbage collection from large institutions. Sanitary considerations make it obligatory to cook the garbage before feeding, introducing the possibility that availability of some nutrients may be reduced by the heat treatment. A classical study[4] concluded that one ton of unsupplemented residential garbage would produce 39-80 pounds pork. It was shown that this could be increased if the garbage was fed with an appropriate supplement.

In a systematic evaluation of garbage as a swine feed the proximate composition of garbage, and the extent of its variation was measured[5]. Samples were collected on different days of the week and different seasons of the year from four different sources: hotels and restaurants, institutions, military, and general municipal collection. The results of this study are shown in Table 2.

There was a two-fold difference between the dry matter contents of Institutional and Military garbage, but the average protein level in the garbage from the four sources was quite uniform and would be adequate to supply the needs of the pig. The large coefficients of variation indicate, however, that many individual samples would have contained protein levels below the pigs' requirements. The fat contents were all greater than the amount which would be supplied in cereal based diets. All except the Institutional garbage would have been sufficiently high in fat to reduce (or even eliminate) endogenous fat synthesis. This may account for the discrimination against garbage fed pigs which is often exercised by meat packers. The dietary fat deposited directly into pig

Table 2. Proximate Composition of Garbage from Four Different Sources in New Jersey. Values based upon analysis of 20-30 samples from each category.

	Dry Matter (% of Cooked Product)	Crude Protein	Crude Fat	Crude Fibre	Ash
		(% of Dry Matter in Cooked Product)			
	Hotel and Restaurant				
Mean	16	15	25	3.3	5.7
Coefficient of Variation	28	24	33	43	23
	Institutional				
Mean	14	14	12	2.8	4.2
Coefficient of Variation	24	23	57	53	18
	Military Establishments				
Mean	28	16	34	2.9	5.6
Coefficient of Variation	24	23	31	49	29
	Municipal Collection				
Mean	17	18	21	8.4	8.6
Coefficient of Variation	46	26	34	54	44

carcasses from garbage is often softer and more oily than the fat synthesized by the pigs from dietary carbohydrate supplied in typical cereal based diets. Day of the week upon which garbage was collected had no influence upon its proximate composition. Season of the year only influenced crude fibre and ash levels; these were slightly higher in summer than in other seasons.

Calcium and phosphorus contents of garbage were either less than or barely adequate to meet the pigs' needs for these minerals. Thus, pigs responded to supplementation of garbage diets with calcium and phosphorus by increasing rate of growth and improving bone development[6]. Assessment of the vitamin concentration of garbage in relation to pigs' requirements showed that all the garbage samples analyzed were deficient in pantothenic acid. Some samples contained only marginal amounts of carotene and riboflavin[7]. However, even when a garbage diet is supplemented with nutrients which are likely to be deficient, pigs still do not grow as rapidly as pigs which receive a protein supplemented, cereal diet. It is concluded that the high water content of the garbage limits amounts of nutrients which can be consumed by the pig because of the sheer bulk of the liquid diet. Successful pig production based upon the use of garbage requires the use of dry feeds in addition to the garbage to allow intake of adequate quantities of nutrients to support normal growth rates.

Wastes with Potential as Feed Resources

Sugar production results in the generation of large quantities of molasses. These by-products or 'final molasses' are characterized by high ash content (10%) and contain sugars as the major organic constituents (sucrose 40%, glucose 10% and fructose 10%). They are practically devoid of protein. The possibility of using this material as a feed energy resource for meat production has inspired many investigations, particularly in tropical countries where cereal production is not agronomically attractive. The problem is that substitution of more than 40% of final molasses for cereal in the diet results in diarrhea in pigs and reduced growth rates[8].

The coincidence that growth of pigs receiving garbage diets is limited by inability to consume enough nutrients and that use of molasses is limited because of the pigs' inability to digest diets with more than 40% molasses has been exploited by Cuban animal scientists. A feed mixture of 60% garbage and 40% molasses fed with a dry protein supplement offsets the high water content of the garbage with the low water content of the molasses. This allows pigs to consume adequate amounts of energy for near maximum growth. The fibre and other organic constituents of the garbage dilute the molasses in the diet promoting the digestion of the energy yielding sugars. The absence of protein in molasses is offset by provision of protein in the dry supplement which is added to the garbage: molasses mixture. Such a system allows production of pig meat using garbage which would otherwise pose a disposal problem and molasses products which are plentiful and relatively cheap in tropical, sugar-producing countries.

There is continuing interest in the exploitation of feed resources which are relatively rich in protein. Meat meal and fish meal are well established as protein feeds. Meat packers also produce large volumes of keratinized protein residues of animals; hair and feathers. These materials are very resistant to biological degradation and for many years this stability has enabled their use in upholstered products. This outlet has largely been replaced by synthetic materials making disposal of the keratin a potential economic problem. However their stability could be disrupted by the destruction of the disulfide bonds[9]. This can be accomplished by heating. The heated product was well digested by swine and poultry. However, destruction of the disulfide bonds results in a product which is deficient in sulfur amino acids. Thus, although hydrolyzed feather meal and hydrolyzed hog hair have high protein levels, they must be used in carefully balanced diets for pig and poultry production[10].

Microorganisms are being used to convert organic wastes to microbial mass; such products are referred to as 'single cell protein' since the microbial mass is predominantly protein. The wastes usually provide a source of carbon compounds which can be oxidized by microorganisms to support their energy metabolism. Nitrogen and other mineral elements needed for microbial growth are provided as inorganic salts. Great efforts are being expended upon the use of hydrocarbon residues as energy sources or fuels for the microorganisms. These are the short-chain volatile compounds which have traditionally been burnt off as by-products of hydrocarbon processing. Other wastes used as substrates are those arising from the pulp and paper industry. Disposal of these wastes has caused serious deterioration of natural water courses and lakes. Yeast has always been a by-product of brewing but alcoholic fermentation is carried out under anaerobic conditions which limit the growth of the microorganisms. If the yeast are allowed to grow under aerobic conditions, the carbohydrate substrates are oxidized completely and the production of the yeast is greatly enhanced. A variety of soluble carbohydrates can be used by the yeast and commercial operations using molasses have been established. These units use four tons of molasses to produce one ton of dried yeast. This last fermentation offers the possibility of producing meat using molasses as the major feed resource.

Feces are a troublesome waste of intensive animal agriculture particularly where large concentrations of livestock make impractical or expensive to use the feces as manure on crop land. Poultry manure is recognized as a valuable source of nitrogen which can be used in ruminant diets. However, pig feces do not have this attribute. Research is continuously being carried out to recover some of the nutrients from pig feces. This usually involves some type of fermentation; either aerobic or anaerobic, but it has been found that neither 'dried anaerobic surface residue', or 'oxidation ditch mixed liquor' could be included in swine diets without impairing growth performance in spite of the fact that both products contained an array of the nutritionally important amino acids[11]. An alternative approach is to grow algae on the pig manure, harvest the algae, and use this as an animal feedstuff[12]. There is a species of blue green algae which are easily harvested and contain more than 50% crude protein. This product was well utilized by rats. There are no reports upon its usefulness for pigs. A biologically more sophisticated but technologically simpler system, is to use swine wastes as fertilizer for fish ponds. In many tropical countries, it is common to culture fish, such as carp, in association with pig production units. The fish either consume the pig feces directly, or consume plankton which proliferate as a result of biodegradation of the pig wastes. The fish are harvested for direct human consumption. Even if aesthetic considerations were to make this practice undesirable, the fish could be processed and included in pig diets as a protein supplement. A comparable biological approach, which may have greater potential in temperate climates is to utilize insects which use feces as feed for the larval stage of development. The resulting larvae or pupae can be separated from the feces and used as a feedstuff. All these systems for recovering undigested nutrients from feces involve exploitation of normal biological degradation of organic wastes along with some human intervention to harvest one stage in the food chain which has been established.

Conclusions: The Nutritional and Monetary Value of Wastes

In general the development of meat production has changed from extensive harvesting of land unsuited to cropping with plants which could be directly consumed as foods by man, to the current practice of diverting a substantial part of the cereal grains produced in developed economies to livestock feeding. In current terms, these cereals are both abundant and relatively cheap. On a world scale this is not always true. To extend livestock production, particularly pig and poultry production, exploitation of new feed resources is needed. Such systems need scientific vision and technological ingenuity. However, there is now a well developed basis of biological information which should allow assessment of feed resources against the total resource of organic material available to an economy. The actual composition of this organic material can now be biologically manipulated to make it more appropriate for use as a source of nutrients by domestic animals.

Such technology exists in many cases but its application to commercial production depends upon the prevailing economic conditions. The alternate decisions concerning the extent to which wastes will be exploited as feed resources will be made by entrepreneurs responding to the values placed by society upon animal products produced on these by-products as compared to the conventional ingredients of animal diets and, to an increasing extent, environmental constraints which prevent use of the disposal systems used in the past.

References

1. C. Y. Cho, S. J. Slinger and H. S. Bayley. Computer storage of nutritional data for feedstuffs and its use for calculating nutrient levels in diets. Feed composition, animal nutrient requirements and computerization of diets. Edited by Fonesbeck, Harris and Keale. Utah Agricultural Experiment Station, Utah State University, Logan, Utah (1976).

2. D. M. Baird, J. R. Allison and E. K. Heaton. The energy value for and influences of citrus pulp in finishing diets for swine. J. Anim. Sci. 38, 545 (1976).

3. K. E. Ekstrom, R. H. Grummer and N. J. Benevenga. Effects of a diet containing 40% dried whey on the performance and lactase activities in the small intestine and cecum of Hampshire and Chesterwhite pigs. J. Anim. Sci. 42, 106 (1976).

4. H. Heitman, Jr., C. A. Perry and L. K. Gamboa. Swine feeding experiments with cooked residential garbage. J. Anim. Sci. 15, 1072 (1956).

5. E. T. Kornegay, G. W. Vander Noot, K. M. Barth, W. S. MacGrath, J. G. Welch and E. D. Purkhiser. Nutritive value of garbage as a feed for swine. I. Chemical composition, digestibility and nitrogen utilization of various types of garbage. J. Anim. Sci. 24, 319 (1965).

6. K. M. Barth, G. W. Vander Noot, W. S. MacGrath and E. T. Kornegay. Nutritive value of garbage as a feed for swine. II. Mineral content and supplementation. J. Anim. Sci. 25, 52 (1966).

7. E. T. Kornegay, G. W. Vander Noot, W. S. MacGrath and K. M. Barth. Nutritive value of garbage as a feed for swine. III. Vitamin composition, digestibility and nitrogen utilization of various types. J. Anim. Sci. 27, 1345 (1968).

8. T. R. Preston, N. A. Macleod, L. Lassota, M. B. Willis and M. Velaquez. Sugar cane products as energy sources for pigs. Nature 219, 727 (1968).

9. E. T. Moran, H. S. Bayley and J. D. Summers. Keratins as sources of protein for the growing chick. 3. The metabolizable energy value and amino acid composition of raw and processed hog hair meal with emphasis on cystine destruction with autoclaving. Poultry Sci. 46, 548 (1967).

10. E. T. Kornegay and H. R. Thomas. Evaluation of hydrolyzed hog hair meal as a protein source for swine. J. Anim. Sci. 36, 279 (1973).

11. B. G. Harmon, D. L. Day, D. H. Baker and A. H. Jensen. Nutritive value of aerobically or anaerobically processed swine waste. J. Anim. Sci. 37, 510 (1973).

12. Po Chung, W. G. Pond, J. M. Kingsbury, E. F. Walker and L. Krook. Production and nutritive value of Arthrospira platensis a spiral blue-green algae grown on swine wastes. J. Anim. Sci. 47, 319 (1978).

R. F. Brokken, James K. Whittaker,
Ludwig M. Eisgruber

5. Past, Present and Future Resource Allocation to Livestock Production

Questions related to direct and indirect uses of land in livestock production have become major issues in the search for solutions to the world food problem. Direct allocation involves the use of land for grazing and holding animals. The greater public concern, however, is with indirect allocation, which involves using cropland to produce grain and other feedstuffs, which in turn are fed to livestock. In fact, one of the resolutions adopted at the World Food Conference in Rome urged rich nations to adhere to simpler and less "calorie wasting" food consumption habits. It was argued that large quantities of grain used in livestock and meat production would thereby be released for the benefit of starving people in the third world and, in the long run, land committed to production of feed grains for animals could be reallocated to production of food grains more suitable for human consumption and, supposedly, more "economical" than meat in the human diet. The economics of grain allocation to fattening beef cattle is perhaps the least understood by its many critics and, therefore, special attention will be paid to this issue.

Economic Rationale

The major influences on costs of livestock production are input prices, technology, and government programs. A rise in prices of inputs relative to product prices, will bring about a decrease in the resource allocation to livestock production. This fact was very evident in the early 1970's when high feed grain prices (e.g., corn, sorghum, oats, barley) led to a decline in the use of grain for livestock. Prices of food grains (primarily wheat) are also closely linked to livestock production through this input price mechanism. A rise in the price of wheat relative to feed grains will cause a reallocation of resources from

See Note, page 100.

feed grain to wheat production. This decrease in feed grain production will cause feed grain prices to rise and, hence, resources allocated to livestock production (including both wheat and feed grains) will fall.

New technology also affects livestock resource allocation by enabling producers to decrease production costs. The development of genetic knowledge of animals has enabled many producers to increase the efficiency of feed conversion and, hence, decrease production costs. One of the major influences on U.S. livestock production costs has been technological advance in feed grain production. The development of hybrid corn, for example, greatly reduced the price of corn, and thus led to an increase in grain allocated to livestock. The use of growth stimulants and antibiotics as feed additives increases feedlot gain per pound of feed.

Government programs and restrictions also have a major influence on production costs both directly (such as banning growth hormones in feed and regulation on waste disposal) and indirectly (such as policies affecting input prices). In addition to affecting costs, government policies often affect the revenue from livestock production. Meat import quotas have been used to influence domestic beef prices. The reader is urged to consult an excellent discussion of the historical, socio-economic, technological, and policy development of U.S. agriculture by Earl O. Heady [1].

Other factors that affect revenues from livestock production through their influences on livestock prices are per capita income, population, and tastes and preferences of consumers. Increases in income per capita elicit corresponding increases in consumption for many products. Most meat from livestock (especially beef) falls in this category. An implication of this relationship is that as lesser-developed countries advance economically, the increase in income will create increases in livestock demand and, hence, prices, which will cause corresponding increases in resource allocation to livestock production.

Changes in tastes and preferences of consumers also have a major impact on livestock prices. The recent concern over cholesterol has caused a decrease in the demand and prices for eggs, resulting in a decrease in resource allocation to egg production. Changes in consumer tastes for fat beef have caused a restructuring of prices among beef grades and a corresponding movement of resources from Prime to the lower beef grades. Current concern over excessive caloric intakes of humans and other health issues may influence future resource allocation to livestock.

The argument that beef producers produce what will make them the most profit and foist their products on the consumer who would really rather consume something else is false. In certain industries, where producers are large relative to the size of the market, producers do have the power to "control", to some degree, product characteristics and aggregate output. In the livestock industry, however, each producer is so small relative to the size of the market that any individual's production decisions do not have noticeable effects on aggregate output or prices. The only way agriculture has succeeded in controlling production of widely produced commodities has been through controls established through either mandatory or price-incentive programs sponsored by the government. Enterprises, large or small, are faced with consumer sovereignty to some degree and do not succeed in producing what consumers do not want.

Finally, one of the greatest influences on livestock production has been population growth. Increases in world population have exerted a constant upward pressure on livestock production for both food and power, and for the foreseeable future, this force on livestock demand will likely continue to cause increases in resources allocated to livestock production.

Many noneconomists contend that the economic theory of demand and supply is only relevant for higher income countries or that a different economic theory is needed for lower income countries. This is not true. It is true that producers in some countries may have "non-market" objectives in their livestock production decisions (e.g., breeding cattle to maximize the size of the animal's horns), but economic theory is still applicable. The user of this theory must, however, keep in mind that the concept that producers maximize money profits may be irrelevant in some instances. Instead, producers may maximize personal satisfaction in some form or other, or just try to maintain a subsistence level of production.

World-Wide Production and Consumption

It is well known that as real per capita income rises, the proportion of food made up of meat and animal products also rises. This relationship holds within a country over time as income levels rise and at any given time among families of different income levels. Regier and Goolsby [2] have also shown a surprisingly close association between the levels of per capita income and per capita meat consumption among world regions. Further, Weber and Gregersen [3] and Weber and Weber [4], observed that as soon as income level permits shaping consumption according to the consumer's taste, beef becomes--aside from certain delicacies--the most preferred of all animal protein sources.

Regier and Goolsby [2] estimated the relationships between per capita income and per capita consumption of meat and food grains and per capita use of grain for livestock production with 1962 data from a cross-section of World regions. These relationships are diagrammed in Figure 1. The bottom line, labeled MEAT, shows the relationship between per capita meat consumption and per capita income (real gross national product divided by population). Meat includes beef from cattle and buffalo, veal, all pig meat, poultry meat, lamb, mutton, goat meat, and other meat (including horse meat and game). Fish, eggs, and milk are important sources of animal protein not included in this diagram. Regier and Goolsby [2] note a number of features about this relationship that deserve special attention: In the income range of $80 to $150 a year per capita, meat consumption increases in proportion to income. Below this range, the evidence suggests that meat consumption tends to grow in more than direct proportion to income. This feature seems to reflect the strong drive, at low income levels, to improve the quality of diets. For incomes higher than $200 per capita (1962 dollars), meat consumption continues to rise strongly, but less than proportionally to income. This proportion tends to fall as income rises.

The middle line, labeled GRAIN: FOOD in Figure 1, shows the relationship between per capita consumption of food grains and per capita income. Coarse grains as well as wheat and rice are included in food grains for this relationship. Human per capita consumption of cereal grains averaged 129.2 kilograms throughout the world in 1962 [2]. At very low levels of income, per capita human consumption of cereal grains rose more than proportionally with increases in income. Average per capita consumption among the developing countries was 133.8 kilograms. The relationship shows that per capita consumption of food grains peaks at approximately $100 real per capita income. However, several countries with per capita incomes (i.e., 1962 real gross national product divided by population) in the $500 to $900 range had the highest levels of per capita grain consumption. Factors in addition to per capita income alone must be drawn on to explain food grain intakes for some countries in this per capita income range. The highest consumption rates were in the centrally planned countries of Eastern Europe (155 kilograms) and in the Soviet Union (171 kilograms). Since 1962, these countries have put forth a major effort

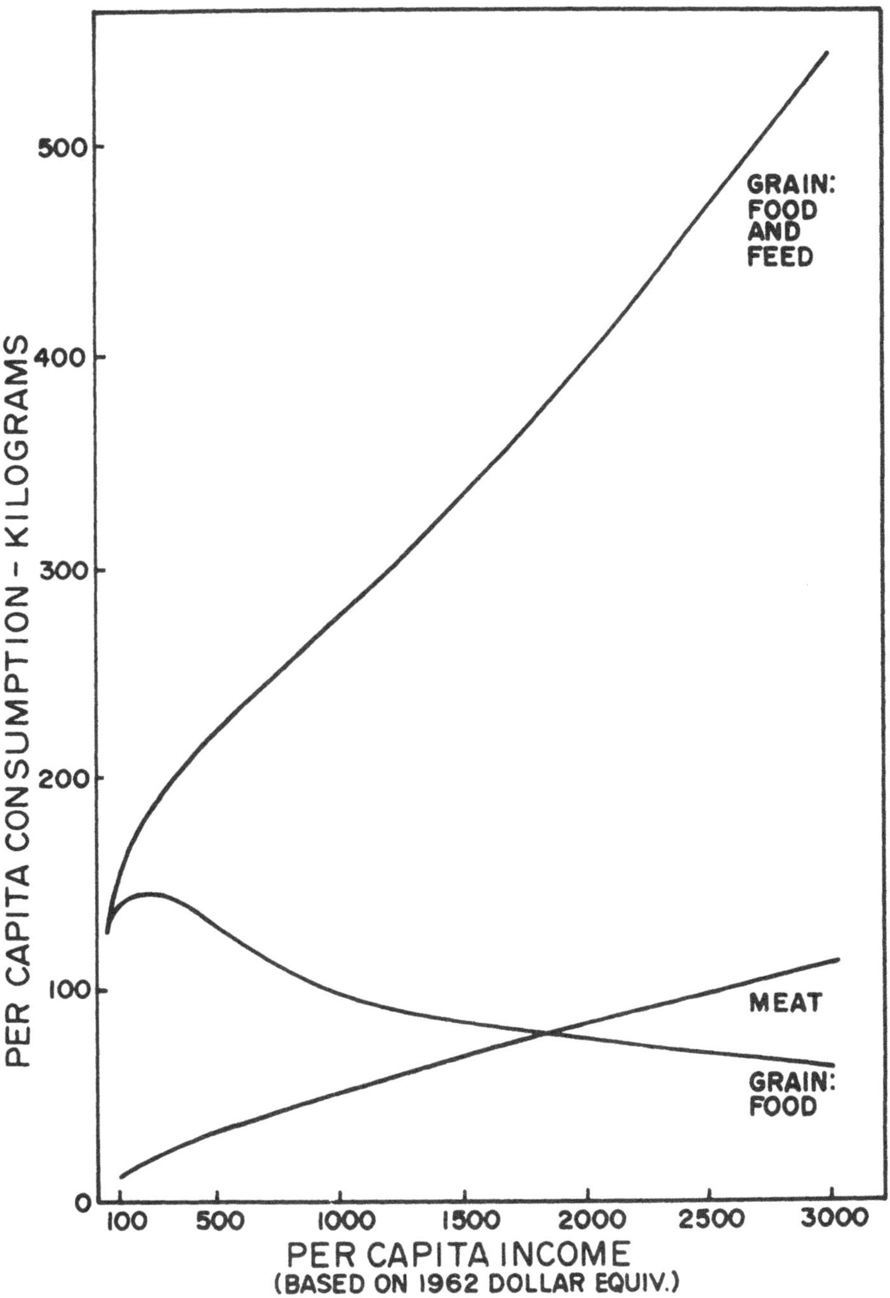

Figure 1. World per capita consumption of meat and grain for food and feed in relation to real per capita gross domestic product--U.S. dollar equivalent. Source: Regier (5) and Regier and Goolsby (2).

to develop their livestock economies. In times past, the Soviet Union has tightened its belt and reduced livestock production in times of grain short-falls and have maintained relatively high consumption of food grains during these times. More recently they have chosen to increase imports of grain in order not to suffer a setback in the substantial gains in livestock production. The non-communist higher income regions of the world fell very close to the fitted line. Their average annual cereal consumption was 98 kilograms.

Perhaps the most dramatic of the three relationships shown in Figure 1 is the relationship between use of grains for livestock feed and human food and per capita income. This is the top line in Figure 1. The difference between the top and middle lines represents the relationship between real per capita income and per capita use of grain for livestock feed.

This relationship between grain use per capita and per capita income is the basis for the proposition that simpler food consumption habits in the rich countries of the world could help alleviate hunger during times of food emergencies in the third world. Such a tactic requires that livestock must exist in the first place so that a food consumption base from which to cut back exists. In fact, livestock have played this role in the past when world food grain reserves were inadequate to meet an emergency. The reallocations to food are accomplished in the market place by bidding grain prices too high for use as feed grains. This causes havoc for livestock farmers, and is most unpleasant for consumers, government planners and aid recipients alike. An international food grain reserve program could minimize reliance on the livestock sectors as secondary grain reserves.

The relationship between per capita use of grain for livestock and per capita income also suggests that rising per capita incomes in the developing nations will exert strong pressures on world agricultural resources on one hand, but also suggests that rising incomes provide economic incentives for further development of agricultural capacity. Thus, in addition to developing a primary food reserve in the form of livestock inventories which can be slaughtered in food emergencies, there are also secondary grain reserves in the form of feed grown for livestock that can be diverted to human food. Thus, the development of a livestock sector enlarges the production base from which food can be diverted to human needs in times of food emergencies. The economic disruptions from such periodic emergencies can be severe. Again, reliance on such secondary reserves could be reduced through development of a world food grain reserve.

Several analysts have noted the potential for livestock as a secondary grain reserve [5,6,7]. In his analysis of the role of livestock in alternative futures in the world grain, oilseed, and livestock economy, Regier [5] found that production, consumption, and trade of livestock products was quite sensitive to grain prices among world regions. This balancing role of the livestock sector has been incorporated into recent USDA projections, for 1985, of regional commodity balances in the world feed-livestock economy [8].

In 1974-75, total concentrates (grain, grain by-products, oilseeds and oilseed meals) fed to livestock in the United States dropped 42 million metric tons (mmt) or 22 percent. All grains fed dropped 33 mmt or 24 percent [9]. Concentrates fed to feedlot cattle dropped 26 mmt or 49 percent. Thus, admonitions to feed less grain to cattle were substantially carried out by the market place. It is worth noting, however, that in addition to the incentive to feed less grain (provided by the sharp increase in grain prices) there was also a sharp drop in beef prices and meat prices generally, due to very large livestock inventories.

Fitzhugh et al. [10] see an important role for ruminants (cattle, sheep, goats, water buffalo, camels) in the future. They note that 75 percent of the world's agricultural land produces forage of one kind or another that can be utilized only by ruminants. They point out that, with the exception of India and certain countries in North Africa and the Middle East, forage resources for ruminants are far from fully utilized and can support considerable expansion within the present forage resource base in most regions of the world. Furthermore, considerable expansion in the land base and its forage productivity can also be attained in most regions of the world. They estimate that approximately 15 percent of the total increase in feed resources (for ruminants) would derive from grain on a metabolizable energy basis.

Milk and meat from grain consuming animals provided 69 percent of the world's consumption of animal protein during the mid-1960's [3]. Meat from grass consuming animals includes mostly that from domesticated ruminants, but also includes some meat from horses and wild game. In addition, ruminants provide as much as 99 percent of power for agriculture in the developing world.

It is extremely important to note that only 5 to 10 percent of feed consumed by ruminants world-wide is

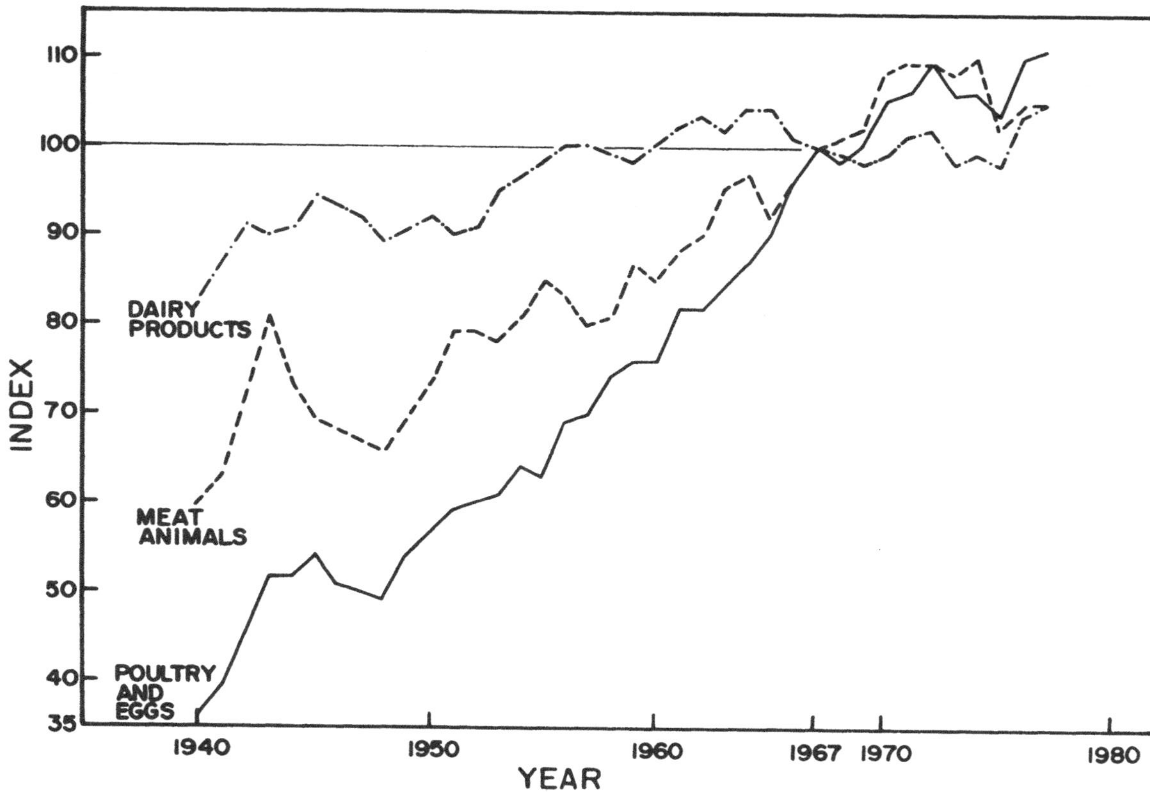

Figure 2. Index of total U.S. output (1967 = 100) of dairy products, meat animals, and poultry and eggs, 1940-1977 (12).

potentially consumable by humans [10]. Even in the United States consumption of grain and by-product feeds constitutes only about 30 percent of the nutrients used by cattle and sheep. Cattle for milk and beef production used 40 to 45 percent of concentrates fed to all livestock in the United States. Only about 20 percent of nutrients fed to beef cattle comes from grain and other concentrate feeds. Moreover, without sacrificing numbers fed, feedlot cattle can be fed to present slaughter grades on much less grain by first carrying them to heavier weights on pasture or range and by substituting some harvested roughages for grain in the feedlot stations. However, given present grain prices, this can be done only at higher cost. Grain and by-products constitute about 92 percent of the nutrients fed to pigs and 97 percent of the nutrients fed to poultry in the United States.

Production, Consumption and Resource Use in the U.S.

In this section, some trends pertaining to production, consumption, prices of livestock products, and resource use by livestock in the U.S. are presented.

Figure 2 shows indices of total U.S. output of dairy products, meat animals (cattle and calves, sheep and lambs, and hogs), poultry and eggs from 1940 to 1977 [12]. Output of poultry and eggs has shown the largest percentage increase over this period. From 1940 to 1977, poultry and egg production rose 197 percent. During this same time period, production of dairy products rose 27 percent, and production of meat animals increased 75 percent.

Meat consumption per capita has also increased a great deal since 1940. It has risen 57 percent. A breakdown of aggregate meat consumption into beef, pork, and poultry in pounds of carcass equivalent per capita is shown in Figure 3. Per capita beef consumption has risen 130 percent since 1951. During the same period, per capita consumption of poultry rose 101 percent while pork consumption dropped 19 percent [9].

Two major factors have caused these large increases in U.S. per capita meat consumption. The first is a steady rise in real per capita disposable income from $1,367 in 1940 to $3,233 in 1977. This increase of 136 percent in disposable income over the past 38 years has tended to increase total meat consumption and the proportion of meat consumed that is beef. The second factor that caused increased U.S. meat consumption is that while income was rising, real retail prices decreased. By real prices, we

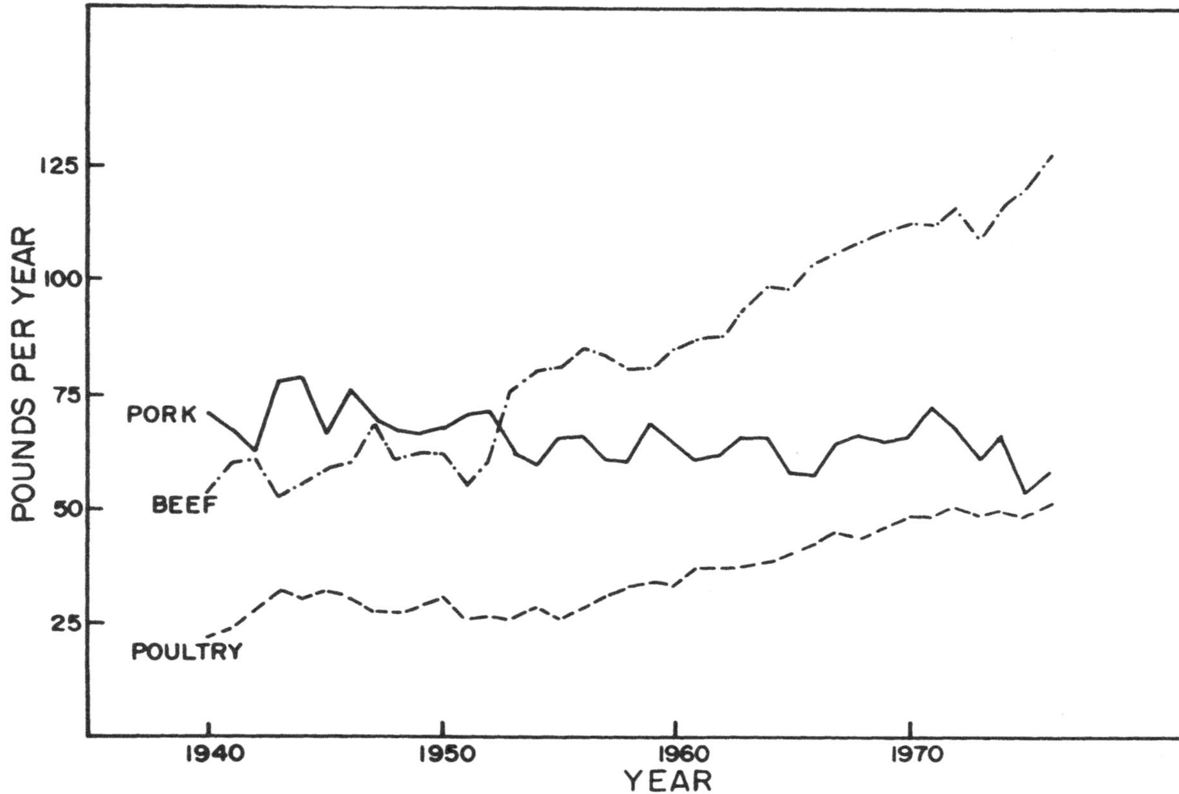

Figure 3. Annual per capita consumptions of beef, pork and poultry in pounds carcass equivalent weight (9).

mean actual prices adjusted for changes in the value of the dollar over time. We used the retail price index for all items: 1967 = 100 as a basis for adjusting the prices to a real dollar value basis. Respectively, real prices of pork, beef, and poultry have fallen 7, 23, and 58 percent. The decreases in prices of beef and poultry relative to pork have also contributed to the rapid increases in consumption of beef and poultry relative to pork.

The reason U.S. Livestock production increased rapidly (simultaneously with decreases in livestock prices) is the large increase in resource productivity that has occurred in U.S. livestock and grain production. Figure 4 shows the index of total farm labor used for livestock production and the index of output of all livestock and livestock products from 1940 to 1977. The drop in labor used in livestock production is quite dramatic. In 1943, the index was 246, but in 1977 it was 56, less than one-fourth the earlier level. In the same time period, livestock output increased 38 percent. Output per man hour has more than doubled in the last twelve years for milk and poultry, and has increased 72 percent for beef and pork.

Figure 5 shows another example of the increase in productivity in the U.S. livestock sector. This figure contains indices of acreage used in the production of feed grains used in livestock production and of output of livestock and livestock products. The acreage index peaked in 1943 and since then has decreased over 50 percent, but livestock production has increased steadily. This dramatic change was made possible by increased yields per acre and increased feed efficiency in livestock.

Real grain prices have fluctuated rather wildly at times, but the main trend has been dramatically downward. Following World War II, the real price of corn peaked at \$6.43 per bushel (this was the overall U.S. average price for the year during 1947). Then the price dropped to \$3.48 per bushel in 1948 and has since trended generally downward to \$2.04 per bushel in 1967. Then rather wide price fluctuations again set in. The weighted average price for the year reached \$3.66 per bushel in 1974, but in 1977 it reached a new all time low of \$1.12 per bushel (in real prices, 1967 = 100). The current real corn price is even lower.

Some Misconceptions

There are two popular misconceptions about the economy of feeding grain to cattle and feeding cattle to heavy weights. The first misconception is that grain is a more

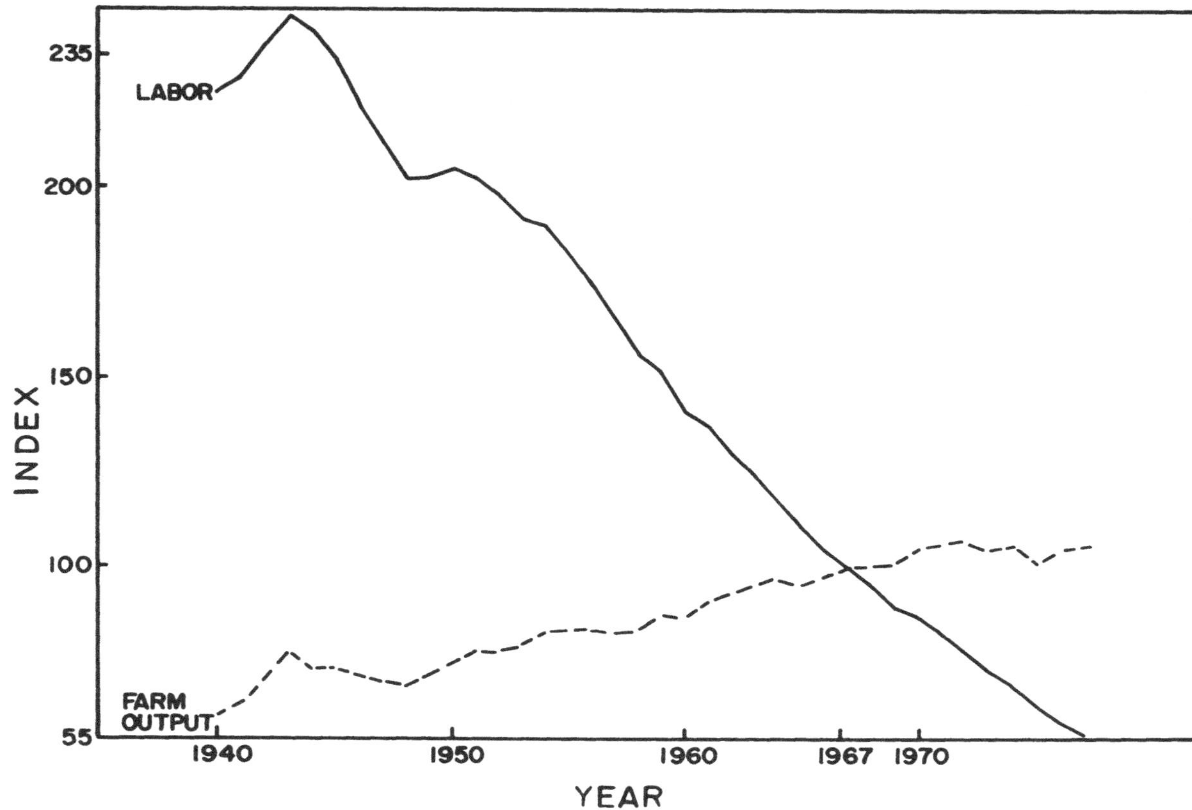

Figure 4. Index of total labor hours used and total farm output of all livestock and livestock products, 1940-1977 (12).

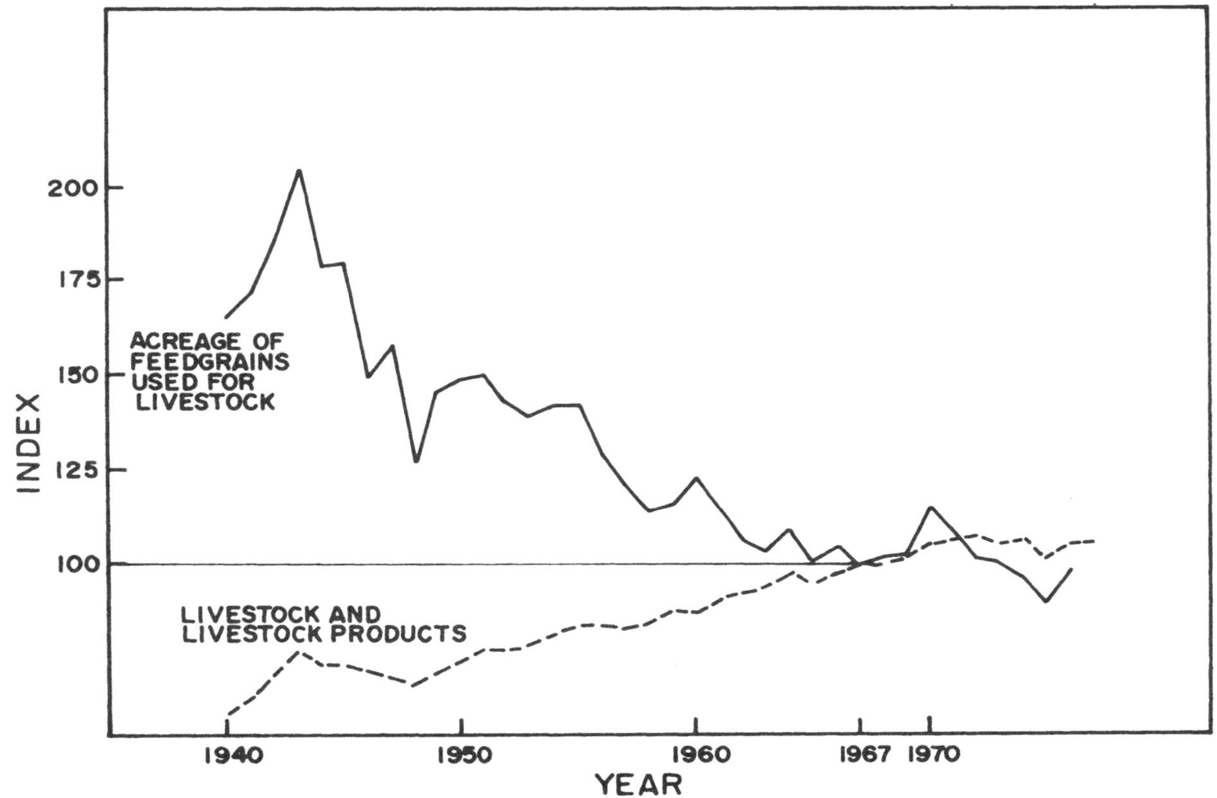

Figure 5. Index of acreage used to produce feed grains used for all livestock and index of total United States output of livestock and livestock products, 1940–1977 (1967 = 100) (9, 12).

Table 1. Cost Per Calorie of Alfalfa, Corn and Silage in Terms of Net Energy for Maintenance (NE_m) and gain (NE_g) in Growing and Finishing Cattle [a].

Year	NE_g Basis			NE_g Basis		
	Alfalfa Hay	Corn	Corn Silage	Alfalfa Hay	Corn	Corn Silage
	- - - - - - - - - -	- - - - - - - - - -	- - - - - c/Mcal	- - - - - - - - - -	- - - - - - - - - -	- - - - - - - - - -
1950	2.47	2.83	2.77	5.19	4.36	4.36
1955	2.07	2.94	2.85	4.36	4.54	4.50
1960	2.17	2.14	2.25	4.57	3.29	3.56
1965	2.12	2.48	2.51	4.46	3.83	3.96
1966	2.32	2.63	2.61	4.88	4.03	4.12
1967	2.47	2.29	2.37	5.19	3.53	3.74
1968	2.42	2.15	2.27	5.09	3.32	3.59
1969	2.32	2.25	2.27	5.09	3.32	3.59
1970	2.37	2.50	2.53	4.98	3.86	3.99
1971	2.76	2.21	2.32	5.81	3.41	3.65
1972	3.22	2.55	2.58	6.75	3.95	4.05
1973	4.15	4.92	4.30	8.72	7.59	6.77
1974	4.79	6.49	5.43	10.07	10.01	8.57
1975	5.18	5.45	4.67	10.90	8.40	7.37
1976	5.44	4.75	4.17	11.42	7.32	6.57
1977	5.98	3.26	3.08	12.56	5.02	4.85
1978	4.54	4.05	3.67	9.55	6.25	5.77

a/ Energy values from N.A.S. [13] costs based on prices received by farmers in Colorado near harvest time [11], alfalfa, August; corn, October; silage price per ton is 6 bushels of corn plus $3.00.

expensive source of feed for cattle than harvested roughages. The second misconception is that beef might become less expensive if people preferred lower grading (less fatty) beef.

Energy required per pound of gain decreases as daily rate of gain increases and the animal's rate of gain increases as the proportion of grain in its diet increases [13,14,15,16]. Therefore, fewer calories are required per pound of gain as the proportion of grain in the ration is increased. Moreover, the cost per calorie of feed energy is often lower from grain than from harvested roughages. Costs per calorie from corn, alfalfa hay, and corn silage are shown in Table 1. These costs were calculated using prices received by farmers in Colorado on a net energy for maintenance (NE_m) and net energy for gain (NE_g) basis [13]. This is the energy system commonly used in formulating feedlot rations.

According to this system, roughages are relatively more valuable in feeding for maintenance such as in maintaining adult animals at zero weight gain than in feeding for gain. On the basis of relative energy values for weight gain, corn is less costly than alfalfa hay in every year shown in Table 1 except 1955. Until 1973, corn was generally lower in cost per calorie of NE_g than corn silage. Since then, corn silage is shown to be less costly per calorie than corn grain. The method used to calculate the silage prices probably is biased downward for the more recent years (see footnote a, Table 1). Even so, rate of gain is higher on high energy rations consisting mostly of grain, than on all silage rations. This is especially true in the latter phase of the finishing period. Therefore, many feeding programs start on silage and phase into high energy, grain based rations for the final stage of finishing.

Throughout most of the last three decades, cattle feeders have paid more per pound for feeder cattle than they received per pound for their fattened cattle. For example, the report on feeder cattle sales at Omaha, Nebraska for the week ending December 2, 1978 was \$76.25 per cwt [17]. Quotations for 900 to 1,000 pound Choice beef were \$55.38 for the same week. The June futures price for Choice beef on December 1, 1978 was \$62.65 per cwt.

Suppose a cattle feeder purchased a 500-pound steer for \$76.00/cwt or \$380 and sold it at 1,000 pounds for \$63.00/cwt or \$630. With no death loss, he must produce the 500 pounds gain for \$250 or 50 cents per pound to break even. This kind of feeding program is now taking place and indicates clearly the strong role feedlots play in maintaining

Table 2. Approximate Feedlot Production Costs for Steer Calves with Corn @ $2.25/ bushel.

Weight	Gain	Days on Feed	Total[a] Feed Consumed	Feed[b] Cost	Other[c] Costs	Total Costs	Total Costs ÷ Total Weight	Cost of[d] Last lb.
lb	lb/day	days	lb	$	$	$	$/cwt	¢/lb
450					270.00	270.00	60.00	
450	1.98	19	149	7.17	305.63	312.80	69.51	28.91
500	2.09	44	372	17.95	308.49	326.44	65.29	27.96
600	2.26	89	857	41.37	313.64	355.01	59.17	30.01
700	2.37	132	1402	67.69	318.57	386.26	55.18	32.35
800	2.44	174	2009	97.00	323.38	420.38	52.55	34.89
900	2.45	215	2666	128.77	328.07	456.84	50.76	37.59
1000	2.42	256	3379	163.22	332.77	495.99	49.60	40.52
1050	2.39	277	3763	181.73	335.17	516.90	49.23	42.12

[a] Feed consumption and rates of gain based on example given in "Nutrient Requirements of Beef Cattle," N.A.S. 1970 Ration Energy Content: NE_m = 2.16 Mcal/kg, NE_g = 1.4 Mcal/kg. Based on 84% corn, 10% corn silage, .045% soybean meal, 1.5% other, 100% dry matter basis.

[b] Feed prices: corn $2.25/bu; silage @ $16.50; soybean meal @ $190/ton.

[c] Includes purchase cost, 4¢/day yardage, $3.00/ton feed markup, $8.55 initial treatments and acquisition costs, $23.55 vet and other, and interest on purchase costs and on $8.55 at 9%.

[d] Includes feed, yardage, profit, and interest on purchase price and on initial start-up costs.

supplies and keeping beef prices from rising faster than they might without active role of feedlot production.

It is important to understand the difference between average cost per pound of a growing animal and its marginal cost. Average cost is total cost accumulated over the life of the animal divided by its current weight. Marginal cost is the cost of the last pound of growth. Average cost per pound of an animal is highest at birth; marginal cost is never lower than at birth. As the animal grows, marginal cost rises, while average cost declines. Average cost continues to decline as long as marginal cost remains below average cost. Thereafter, average cost rises.

The total, average (total cost ÷ total weight), and marginal costs (costs of last pound) at different weights as an animal grows are shown in Table 2 for an animal started in the feedlot at 450 pounds and in Table 3 for an animal started in the feedlot at 700 pounds.

In neither case does average cost reach a minimum before the animal reaches usual market weight. This relationship is most significant because it shows that the preference for Choice beef in the United States has not been inconsistent with the goal of producing beef at the lowest cost per pound. In fact, over the past three decades, a preference for a lower degree of fattness would have resulted in higher average prices for beef than we have experienced.

This conclusion is born out by other methods of analysis as well. For example, Ward et al. [18,19,20] analysed resource requirements of alternative beef production systems with particular focus on minimizing production costs vs. minimizing energy use. They found that even when they assumed that consumers were indifferent to the quality on grades of beef, the least cost system did not differ appreciably from present practices, i.e., producing Choice beef with grain. Moreover, the main avenue to reducing energy use was not in feeding less grain to cattle, but in relocating where they were fed so that the use of irrigated grain for cattle was reduced.

Brokken et al. [20] analysed the national beef production system with respect to effects of reducing fat content by marketing animals at lower weights. This analysis demonstrated that heavier market weights yield the lowest production cost per pound (in terms of retail cost per pound). This is not too surprising to those who have studied the cost of maintaining a cow herd or the per pound cost of producing a live baby calf.

Table 3. Approximate Feedlot Production Costs for Yearling Steers with Corn @ $2.25/bushel.

Weight	Gain	Days on Feed	Total[a/] Feed Consumed	Feed[b/] Cost	Other Costs	Total Costs	Total Costs÷ Total Weight	Costs of[d/] Last lb.
lb	lb/day	days	lb	$	$	$	$/cwt	¢/lb
700					350.00	350.00	50.00	
700	2.84	15	199	9.61	388.41	398.02	56.86	30.88
800	2.89	50	753	36.38	393.11	429.49	53.69	33.27
900	2.89	85	1401	66.77	397.80	464.57	51.62	35.92
1000	2.87	120	2064	99.73	402.50	502.23	50.22	38.65
1100	2.81	155	2792	134.87	407.19	542.06	49.29	41.47
1150	2.76	173	3181	153.65	409.61	563.26	48.98	42.96

a/ Feed consumption and rates of gain based on example given in "Nutrient Requirements of Beef Cattle," N.A.S. 1970 Ration Energy Content: NE_m = 2.16 Mcal/kg, NE_g = 1.4 Mcal/kg. Based on 84% corn, 10% corn silage, .045% soybean meal, 1.5% other, 100% dry matter basis.

b/ Feed prices: corn $2.25/bu; silage @ $16.50, soybean meal @ $190/ton.

c/ Includes purchase cost, 4¢/day yardage, $3.00/ton feed markup, $8.55 initial treatments and acquisition costs, $23.55 vet and other, and interest on purchase costs and on $8.55 at 9%.

d/ Includes feed, yardage, profit, and interest on purchase price and on initial start-up costs.

To produce the same amount of beef per year with lighter animals, or with production systems that require more time to reach a given market weight, requires more of all age and sex categories of cattle. It turns out that these are not attractive alternatives in dealing with inflation in beef prices.

At the present time the beef cow (adult female bovine) inventory is quite low. To grow their offspring in production systems that require extending their life on earth (i.e., to grow them on extensive forage systems) or to market them at lighter than usual weights, would reduce current annual rates of beef production and aggravate the already high beef prices. Fortunately for consumers, corn prices are relatively low. To get the highest rate of production from the low cow inventory, growth rates must be accelerated on as high a proportion of the annual calf crop as possible. This can be done economically when grain prices are low relative to beef prices by placing the calves in feedlots at relatively low weights and growing them at accelerated growth rates on grain. This is exactly what is now starting to take place.

Summary

We have presented the general economic rationale underlying resource allocation to livestock production. Resource and commodity prices are formed in their respective markets through the interplay of the forces of supply and demand. Legal or policy restrictions imposed on either supply or demand also play a role in setting resource and commodity prices. Major influences on the supply side are general availability of resources relative to demand for them (i.e., resource prices), technology, and government programs. Major influences on the demand side are per capita income, population levels, and consumer tastes and preferences.

A review of data from a cross section of world regions demonstrates a strong positive relationship between per capita income and meat consumption and between per capita income and use of grain for livestock feed. Historically, as countries advance economically, per capita consumption of meat increases and as soon as income level permits shaping consumption according to consumer's taste, beef becomes the most preferred of all animal proteins (aside from certain delicacies) [11].

In the United States, rapid technological change has allowed livestock production to increase while total resource use in this production has decreased dramatically. There are widely held misconceptions: that grain is a costly source of food energy for beef and, that feeding beef to traditional grades and relatively heavy weights is a costly and wasteful practice. It is demonstrated that neither proposition is true.

Note

The authors are, respectively, Agricultural Economist, U.S. Department of Agriculture, Assistant Professor of Agricultural and Resource Economics, Oregon State University, and Professor and Head, Department of Agricultural and Resource Economics, Oregon State University, Corvallis, Oregon 97331.
All opinions are those of the authors and not necessarily those of the U.S. Department of Agriculture or Oregon State University.

References

1. Earl O. Heady, "The Agriculture of the U.S.," Scientific American, 23(3):106-27, September 1976.

2. Donald W. Regier, and O. Halbert Goolsby, "Growth in World Demand for Feed Grains: Related to Meat and Livestock Products and Human Consumption of Grain," U.S. Dept. of Agr. Econ. Res. Serv. Foreign Econ. Report 63, July 1970.

3. Adolf Weber and Marquard Gregersen, "Production in the World Cattle Industry, An Analysis of Potentials and Constraints," unpublished draft.

4. ______________, and Ernest Weber, "The Structure of World Protein Consumption and Future Nitrogen Requirements," Eur. R. Agr. Econ. 2(2):169-92.

5. Donald W. Regier, "Livestock and Derived Feed Demand in the World Gol Model," Foreign Demand and Competition Division, Economics, Statistics, and Cooperatives Service, U.S. Dept. of Agr. Foreign Agr. Econ. Report 152, September 1978.

6. L. Soth, "Ruminants: A Flexible Food Reserve," Agr. World 20(1978):6-7.

7. Peter Svedbert, "World Food Sufficiency and Meat Consumption," Amer. J. of Agr. Econ. 60(4):661-6, November 1978.

8. Anthony Rojko, Donald Regier, Patrick O'Brien, Arthur Coffing, and Linda Bailey, "Alternative Futures for World Food in 1985: Vol 1, World Gol Model Analytical Report," Foreign Demand and Competition Division, Economics, Statistics, and Cooperatives Service, U.S. Dept. of Agr. Foreign Agr. Econ. Report 146, May 1978.

9. U.S. Department of Agriculture, "Agricultural Statistics," U.S. Government Printing Office, Washington DC, various issues 1952-1977.

10. H. A. Fitzhugh, H. J. Hodgson, O. J. Scoville, Thanh D. Nguyen, and T. C. Byerly, "The Role of Ruminants in Support of Man," Winrock International Livestock Research and Training Center, Morrilton, Arkansas, April 1978.

11. U.S. Department of Agriculture, "Agricultural Prices," Statistical Reporting Service, Washington DC, Published Monthly 1950-1978.

12. Donald D. Durost, and Evelyn T. Black, "Changes in Farm Production and Efficiency, 1977," U.S.Dept. of Agr. Econ. Stat. Coop. Serv. Stat. Bul. 612.

13. National Academy of Sciences, "Nutrient Requirements of Beef Cattle," Fourth Revised Edition, National Research Council, Washington DC, 1970.

14. Ray F. Brokken, "Economics of Grain-Roughages Substitution in the Beef Sector," Forage-Fed Beef: Production and Marketing Alternatives in the South, Bul. 220, Southern Cooperatives Series, June 1977.

15. ______________, "Effects of Ration Nutrient Concentration on Voluntary Feed Intake and Animal Performance: An Analytic Framework," In proceedings of First International Symposium on Feed Composition, Animal Nutrient Requirements, and Computerization of Diets, Logan, Utah. Utah State University Exp. Sta., Logan, Utah, 11-16 July 1976, pp. 484-90.

16. ______________, T. M. Hammonds, D. A. Dinius, and John Valpey, "A Framework for Economic Analysis of Grain Versus Harvested Roughage for Feedlot Cattle," Amer. J. Agr. Econ. 53:245.

17. U.S. Department of Agriculture, "Market News," Livestock, Poultry, Grain, and Seed Division, Agr. Marketing Serv., 46(48), December 5, 1978.

18. Gerald M. Ward, "Energy and Resource Requirements for Beef Cattle Production Feedstuffs," 48(16), December 1976.

19. ______________, "Resource Requirements for Alternate Beef Production Systems," Executive Summary, NSF Grant No. SIA 75-14125, September 1978.

20. ______________, P. L. Knox and B. W. Hobson, "Beef Production Options and Requirements for Fossil Fuel," Science, 198(4314):265-71, October 21, 1977.

21. Ray F. Brokken, Carl W. O'Connor, and Thomas L. Nordblom, "Analysis of Alternative Beef Growing and Finishing Systems With Focus on Alternative Market Weights," manuscript, 1979, (to be published).

Robert E. McDowell

6. The Role of Animals in Developing Countries

Abstract

The Developing Countries (DC) hold over 60 percent of global domestic livestock units but these provide only 18 to 20 percent of the world supply of meat, milk and other edible products. Environmental conditions, such as variable rainfall, and genetic potentials of animals are inhibitors to high output but the situation is much more complex. In the DC, animals have many roles besides food; therefore, it is impractical to assess the performance of livestock wholly by U.S. standards. There is high dependence on animals to provide goods and services of a non-food nature, such as traction, fiber, skins, animal excreta for fertilizer or fuel, capital storage and cultural needs. There is often a high level of interdependence of man and animals in the DC. The symbiotic nature of relationships among man and animals must be appreciated in determining the role of animals in the DC. Those who assume animals are serious competitors to man in the subsistence agriculture which exists widely throughout DC tend to overlook this interdependence.

Introduction

The world population of domesticated ruminants (buffaloes, camel, cattle, goats, llama, sheep and yak) numbers approximately 2.8 billion head. In addition there are 0.6 billion swine and 5.3 billion poultry. Those countries classified by the Food and Agriculture Organization as developing countries (DC) hold over 60 percent of these animals. These animals provide no more than 18 to 20 percent of the world supply of edible animal products. Tropical Africa, for example, has one-eighth of the world's cattle but output of meat and milk is low. In this region of Africa it may take up to 200 head of cattle to produce one ton of meat. In contrast, only 20 head are required in Australia and

Argentina and about 5 in the U.S. The reasons for high animal populations and low productivity in the DC are numerous but as will be shown, there is wide variability in the contributions made by animals.

In the U.S., we have come to depend upon milk and milk products, meat and eggs to provide about 67 percent of our total daily intake of protein and 35 percent of the energy we ingest. Our livestock consume approximately 60 percent of our cereal grain production. This is in addition to the nutrients our animals consume as non-edible by-products from the preparation of our food and from grazing our rangelands and pastures.

It would be catastrophic to attempt to extrapolate or transfer our model for animal production to the DC. Cereal grain production in these countries is already either inadequate orbarely so for minimum human needs as food. Based on the usual premise that animals compete with humans for food resources, those living outside the DC often see animal keeping as foolhearty where human needs for food are acute. Those who begin with the assumption that animal populations interfere with human welfare blind themselves to the high degree of interdependence of humans and animals in these subsistence regions. In the DC, there is a symbiotic relationship which is essential to man's existence. Although as yet not fully quantified, we are certain that humans and domestic livestock or fowl enhance each others survival potential and quality of life (1).

Traditional Systems

The order of priorities for farmers keeping livestock in the United States differs in several respects from the usual priorities in the DC as shown in Table 1. Similar to the U.S., the vast majority of the animals are eventually consumed as meat and milk is utilized in the homes but overall, attention is not given to obtaining a high rate of output because other goods and services are more important to the people of the DC than maximizing income from the sale of animal food products.

With farmers in the DC placing high emphasis on reduction of risks or hedge against poor crop yields, they want as large an inventory of animals as possible. This in itself contributes to low performance because there are usually more animals kept than resources will adequately support. The overstocking of animals is partially due to the land tenure systems usual within the DC. Land is often not individually owned and banking institutions are unavailable.

Table 1. General Order of Priorities for Livestock Ownership in the United States and Developing Countries.

U.S.	Developing Countries
Derive income from:	Reduction of risks from cropping
Meat	Accumulation of capital
Milk	Render services e.g.
Eggs	Traction, Fertilizer, Fuel
Fiber	
Diversification of farm operations	Satisfy cultural needs
	Insure status or prestige
Extend use of non-arable lands	Provide food
Prestige	Generate income
Generate capital	

The only opportunity farmers have for investment is in animals. They, therefore, invest their agricultural surpluses in livestock to generate capital. The animals are privately owned and exchangeable.

The infrastructure of roads, railroads and other means of transport and communication is generally far less than desirable in the DC. Low access to urban markets affects the ability of farmers to derive income from animals. This in turn influences the priorities in the roles animals serve, i.e. the emphasis given to maximizing production of edible products versus low product yield complemented by other goods and services. Urban centers of 50,000 or greater in population in the DC generally have four zones from whence supplies of animal products may be drawn (Figure 1). Proximity to the urban market plays a strong part in the role of animals. In Zone I, which lies within or is closely adjacent to the city, are found dairy, poultry and swine enterprises functioning much like the U.S. model with emphasis on output of eggs, milk and meat. Where roadways are limited, Zone II, illustrated in Figure 1, extends approximately 20 km beyond the city boundaries. Farmers in this Zone maintain almost daily contact with the urban center. Thus, they have an

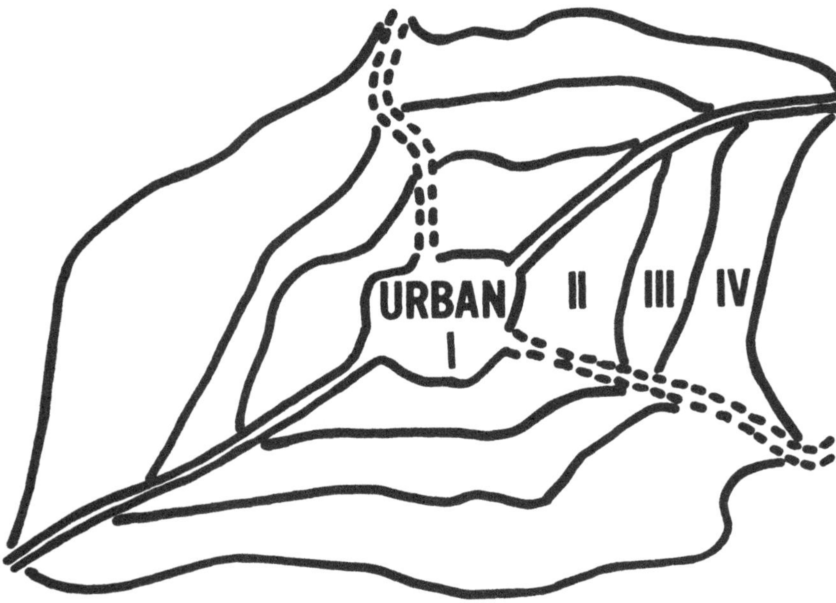

Figure 1. Schematic of the four main zones (I, II, III, IV) for supplies of animal products to cities of 50,000 or more in population as described by McDowell (2). Reprinted with permission from R. E. McDowell, Journal of Animal Science, 46, 1184 (1978).

incentive to produce meat, milk and eggs similar to the farms in Zone I. Zone III ranges between 20 and 60 km or more from the city. Due to the absence of ready transportation, milk, eggs, or animals for slaughter will come from Zone III only when governments or other agencies subsidize collection and transport of milk. No other arrangement is feasible due to limits of refrigerated storage on farms and in households of the city. Animal products must be delivered to consumers within a few hours to avoid spoilage. Some products, such as fermented milk, ghee or cheese, and on occasion eggs or live poultry, may be taken to the city but in this Zone edible animal goods are largely "by-products" of the farming system instead of a routine source of income. Cereal grains or cash crops replace animals as a major source of income. Nevertheless, high emphasis is given to keeping animals for insurance or to hedge against disappointing crop yields. Farms in Zone IV (greater than 60 km from the city) seldom provide fresh milk. Some home prepared milk products may be delivered to the urban center at about monthly intervals. Animals going for slaughter leave the Zone once or twice per year as live animals. Often the animals are moved to market on foot. The trip may last for days or weeks. There is usually little feed available along the way resulting in the animals reaching the market in poor condition. Wool yarn, home crafts using skins or bones and animal hair become the major routine animal products marketed.

Often a large proportion of the animals in the DC are kept by pastoral herders. These herders are found principally in Zone IV (Figure 1) or even further away which means livestock owners seldom have contacts with the urban centers. There are 1.2 million people north of the Sahara in Africa, 6.8 million in West Africa south of the Sahara, 9.3 million in eastern and southern Africa south of the Sahara, 3.4 million in the middle eastern countries, and 1.9 million in central Asia dependent on pastoral herding (3, 4, 5, 6, 7, 8). These 23 million people control over 120 million livestock units.[a] The dependence on pastoral herding varies from country to country but is generally high in the low rainfall areas (600 mm or less per year). Due to inaccessibility of markets and poor returns from crops, the pastoral herders are highly dependent upon their animals for food as well as other means of income, such as the sale of hides, skins, and horns.

People living in the high elevations of the Andes

[a]One animal unit equivalent to one adult cow, or 5 to 7 sheep or goats.

mountains of South America and the Himalaya mountains of Asia also have a high dependence on animals. Less than 5 percent of the land in these regions is cultivated, thus the people expect their animals, yaks or llamas and alpacas, to convert the sparse forage available into human food and produce fiber (hair or wool). Meat, milk, fibers and skins fill essential needs of food and clothing. Animals are also used for transport of goods into and out of these regions. Numerous citations could be given on the multiple ways in which pastoral herders of the arid and semi-arid regions and the people living in the high mountain regions of the world are dependent, in large part, on their herds or flocks of domesticated animals for livelihood. However, this would tend to mask the main point which is that these people rely on animals to provide them with many of their needs for existence.

The roles of animals in the traditional farming or herding systems of the DC are more complex than described; nevertheless, it is readily evident that a very large sector of the human population in the DC have a direct tie with animals.

Food from Animals

Milk. Humans use for direct consumption or as prepared products, such as cheese, approximately 540 million MT of milk per year, but, as is the case with other foods, the distribution of milk is quite disproportionate. Consumption of milk in North America and western Europe, varies from 0.5 to 0.8 liters per day. In the DC daily intake is usually much lower, about 0.3 liters or less than a glass per day. On the other hand, the Masai people of Kenya and other pastoral herders consume 1.0 or more liters per day, while their cattle are grazing during the rainy season. Among certain ethnic groups of southeast Asia, no milk may be consumed beyond early childhood because of human intolerance to lactose in milk from cattle or buffaloes (10). In the U.S., practically all of our milk comes from cows but in the DC buffaloes, goats, sheep, camels, yaks and alpaca are also important sources of milk (approximately 12 percent of world supplies).

From 1965-73, milk supplies increased nearly 2 percent per year in the DC. Growth rate has been less than one percent per year since 1974. One of the major factors responsible for the decrease in growth rate after 1974 was the rise in gasoline prices which added markedly to the cost of moving milk to urban markets. For example, in countries such as Egypt, India and Pakistan, average daily milk production per village farm is about 3.0 liters of which half is consumed by the farm family. With an output per farm of only one to two liters per day, the cost of collecting and moving milk to the urban consumers has become very high.

Desire for milk in the DC is generally high but unfortunately, family income levels are too low to make it attractive for farmers to produce more. As a result, milk supplies will probably increase only slightly in the DC. This will mean a decline in availability per capita due to rapid growth in the human population.

Meat. Protein from various meats, including fish and fowl, is important in most of the DC. Per capita intake of red meat is highly variable, ranging from over 60 kg per year in countries like Paraguay and Uruguay to none in vegetarian societies, such as the Hindu of India. Average consumption of all meats in DC is estimated at 10-18 kg per year with approximately 51% of this coming from beef. The estimates of consumption are probably considerably lower than actual consumption because the use of marine life and meat from animals other than buffalo, cattle, goats or sheep are not included. Frequently, a number of additional species constitute significant sources of animal protein (Table 2). Some of these species provide 25,000 MT or more of meat per year.

Table 2. Some Animals and Fowl Used for Meat in the Developing Countries

Animals		Fowl
Agouti	Goat	Duck
Alpaca	Guinea pig	Guinea
Buffalo	Llama	Pigeon
Camel	Rabbit	Turkey
Capybara	Yak	

In the coastal and high rainfall areas of the DC, fish is an important source of animal protein, but fish production in many areas is declining, especially in irrigated and paddy areas. The use of new high yielding varieties of cereal grains has increased use of chemical fertilizers, pesticides and herbicides. Contamination from the "new cultural practices" has seriously impaired fish production and, in some cases, almost terminated the keeping of ducks which were important local sources of food.

During recent years, poultry production has expanded rather rapidly in the DC, in many countries more than 6 percent per year. This high rate of growth is predicted to continue as poultry products - either eggs or meat - afford a convenient package for family use both in rural and urban areas. For the corresponding years, beef production has increased by approximately 4 percent. The current high cost of fossil fuel may result in some decrease in output of meat from cattle and buffaloes during the next 5 years because of the increased use of these species for agricultural power on small farms. But, over the long range, total meat supplies are expected to rise faster than the human population. Thus, some increase in per capita consumption (10). The inequity of distribution of meat among income groups, especially in urban centers will continue, however.

Even though by U.S. standards the consumption of animal products in the DC is low, small quantities of meat, milk and eggs play a vital role in human nutrition. Often the main human foods are maize, bananas, or root crops, such as cassava, yams or cocoyams. Bananas and the root crops have low protein contents and are especially deficient in the essential amino acid methionine. Where food like cassava is the main diet, a small intake of animal protein is desirable for human health. The edible portion of a two kg chicken, for example, will adequately supplement the protein needs of 8 children.

Non-Food Contributions

When agriculturists from the U.S. or other developed countries attempt to analyze the traditional farming systems of the DC, we fail to recognize that the farmers are frequently trying to have their animals satisfy several objectives; whereas, our production systems are centered on the most efficient method of producing a single product like meat or milk. Classifications of goods and services of non-food nature important in the role of animals in the DC are in Table 3.

Fiber. The main animal fiber in world commerce is wool from sheep. Little of this comes from the DC. This is mainly due to the sheep having a different pelage than usual breeds maintained for wool production. In most DC, sheep have a long hair like fiber ("coarse wool or hair") or hair only. The "long" fibers are preferred as they can be better utilized for home making of yarns than the standard wool fibers. The best carpets of the world are made from the wool of sheep native to the DC. The world's finest grade wool comes from the alpaca and llama in the Andes region of South America. Mohair, which is used to produce the fiber "cashmere", comes from Angora goats that are herded principally on non-arable lands in the arid and semi-arid regions. Approximately 30 percent of global supplies come from the DC.

Table 3. Classification of Goods and Services of Non-Food Nature Contributed by Animals in the Developing Countries

Classification	Some Contributions
Fiber	wool, hair,
Skins	hides, pelts,
Traction	agriculture, cartage, packing, herding, power pumps for irrigation, threshing grains, passenger conveyance
Waste	fertilizer, fuel, methane gas, construction, feed (recycled)
Storage	capital, grains
Conservation	grazing, seed distribution
Pest control	fallow between crops, plants in waterways
Cultural	exhibition, fighting, hunting, racing, status symbol
Inedible products	horns, hooves, bones

Source (1)

Hair from yak of the Himalayas and camels in the arid regions of Africa and Asia constitute a significant source of income for several societies.

Both wool and hair are utilized in cottage industries in more than 100 of the DC for the making of clothing, bedding, housing and carpets. More than 50 percent of the people of southwestern Africa, the Andes region of South America and in the Himalayas of Asia are highly dependent on the sale of hair and wool for handicrafts. Food resources for fiber producing animals come almost entirely from grazing lands too low in rainfall for consistent cropping, or from grazing natural vegetation in mountain regions which are too steep or have elevations too high for crop production.

<u>Skins</u>. The skins of domestic livestock serve as sources of many useful, and for pastoral herders, essential products. They are also a large commodity in world trade. Cattle and buffalo are the main sources of hides, amounting to more than 6 million MT per year. Of this amount, over 40 percent comes from the DC. The finest of leather gloves and handbags are made from the skin of the Red Sokoto goat of West Africa; hence, the primary reason for keeping this type of goat is the value of the skin.

The main value of hides is to produce leather but in certain of the DC, skins are widely used for making household utensils, such as bags for storing water or milk, and in certain areas as foods like <u>Kerupuk kulit</u> in Indonesia, <u>Chickarons</u> in the Philippines, <u>Khak-kwai</u> in Thailand and similar products in Nepal and Afghanistan (11). The export of skins ranks in the first ten commodities earning foreign exchange for about 30 percent of the DC (12).

The value of raw pelts (hides with hair covering) of lambs, goats and camelids is worth more than $3 billion per year. These have a market value as finished products greater than $20 billion. Persian lamb pelts, which rank among the highest in value for producing wearing apparel, come almost entirely from the DC. Sale of these pelts constitutes the primary income for pastoral herders in parts of southern Asia and southwest Africa.

<u>Traction</u>. Traction in several forms (Table 3) is one of the most important roles for animals in the majority of the DC. Even though the use of tractors has expanded, animals are still used to supply 75 to 90 percent of the power used in agriculture. For a large part of southern Asia, there is one draft animal per hectare of cultivated land (13, 14, 15). India is among the largest users of animal power. India would need to expend more than $1 billion annually for fossil

fuel to replace the energy provided by the 70 million bullocks, eight million buffaloes, one million camels and one million horses used for land preparation, threshing of grain and power for irrigation pumps (1). In Thailand and the Philippines approximately 70 percent of the farms use animal power for the preparation of land for planting crops. Numerous countries in central Africa are either introducing or rapidly expanding ox plowing and the use of animal drawn carts in order to expand agricultural production and reduce the drudgery of hand labor (16).

Animals used on farms for draft consume primarily crop residues, such as cereal grain straws; thus, the cost for power to farmers is generally lower than for tractor power. The use of animals for power in agriculture is expected to continue to increase, basically because prices of off-the-farm produced products, like gasoline, and rate of inflation will exceed the rate that the DC governments can condone changes in prices for food grains, meat or milk. These exogenous forces will likely force farmers to resort to more extensive use of animals.

Conservatively, 20 percent of all the world's people have some dependence on the movement of goods by animal cartage or packing (1). For example, anyone in the U.S. enjoying a cup of morning coffee or afternoon tea should recognize that, in all probability, animals contributed to transport of the tea leaves or coffee beans from the mountain areas of tropical countries to the table. Pastoral herders and the peoples of the Andes and Himalaya regions are almost exclusively dependent on animals as pack transport. Animal drawn transportation is vital to the economy of many countries. It provides a vital link to rural villages without roadways and bridges to support trucks. India has 1.5 million animal drawn vehicles (17). Approximately 20 million people are involved part time or full time in the transport of 90 percent of all intra-village goods. About 80 percent of grains are moved from villages to intermediate markets in countries like Egypt and Pakistan. Even in the metropolitan areas of many DC, a high proportion of goods are moved intra-city by animal cartage.

Buffaloes, cattle, horses, mules, donkeys and camels, are employed in various DC for such tasks as riding to control grazing herds or flocks, providing power to move water for irrigation or for threshing of grains by trampling the harvested plants.

Wastes. In the DC, animal wastes serve numerous useful purposes (Table 3). Of most importance to farmers is manure for fertilizer. Its value is so great that often cattle and buffaloes are kept on the farm to provide dung after their usefulness for other purposes has declined. Close to 40 percent of the farmers of the world depend wholly or in part on animal wastes to enhance soil fertility. On small farms, where tillage of land is by hand or with a chisel type plow, farmers prefer manure over chemical fertilizer. Crop farmers frequently pay pastoral herders to herd their cattle on to fields they intend to cultivate at night in order to collect the manure. For instance, in Sri Lanka, sheep are kept on many farms mainly to provide manure. During the day the sheep are grazed on the stubble of the rice fields. At night they are penned on small plots near the house which are used for producing vegetables (18).

More than 200 million MT of manure are used annually as fuel in the DC (1). Dry cow dung will provide 4.6 kcal/gram when burned. With appropriate equipment, dung can prove almost as efficient as coal or oil for cooking. India is the largest user, 60 to 80 million MT annually of dry buffalo or cattle manure is made into dung cakes which are used mainly for cooking. Dung cake fires help control mosquitoes that gather around the household in the evening. In a number of areas of India, up to 60 percent of the total cash income per rural family is derived from the sale of dung cakes. Families in the highlands of Ethiopia realize 30 percent of their income from dung cake sales. This provides a means for women and children to enhance the cash income available to subsistence farms or landless families.

The use of manures for fuel is increasing in many countries due to scarcity of wood and the cost of fossil fuels. Exclusive of distribution costs, India would require over $3 billion per year for coal and oil to replace dung burned as fuel in rural villages and small towns.

Manures can be used in bio-gas plants or digestors to produce methane gas. The gas produced has an energy value of 5 kcal per cubic meter which is 71 percent of the energy value of natural gas (15). In 1975, South Korea had 29,000 bio-gas plants and India 20,000 in villages or on farms. Putting the manure through a digestor does not materially affect its value as fertilizer but the residue is unsuitable for making dung cakes; therefore, farmers accustomed to cash sales for dung cake sales have been reluctant to adopt digestors.

Animal wastes serve numerous other useful purposes. The Masai herders of east Africa plaster the walls of their houses with cow dung. In the highlands of Ethiopia livestock housing is constructed in a similar fashion. A mixture of clay, urine and manure is used as an insect-resistant plaster for the floors of houses in India and other Asian countries. Bricks consisting of cow manure, soil and chopped straw are utilized for house construction in the higher elevations of Peru and Uruguay. In certain cultures, fresh cow manure may be used for poultices as an aid in wound healing.

Storage. In the U.S., a sizeable portion of the grain crop is stored in livestock. For example, animals kept for breeding purposes on U.S. farms are equivalent in calories to more than one year's total food needs for humans. Farmers and others in the DC also use animals to accumulate and store capital, as well as utilizing animals as an emergency food reserve. In total numbers, it is likely that domestic animals and fowl kept for capital storage exceeds all other uses in the DC. Farmers hold on to their animals in order to insure a source of capital for emergencies, such as to buy seed for resumption of farming in case adverse weather causes a crop failure. In the DC of central Africa, farmers regard an increase in animal numbers as the best insurance against economic and social risk as they afford some protection against the uncertainty of rainfall.

Even though poultry and swine make up the largest group maintained for "capital storage", the majority of the goats, significant portions of sheep, cattle, buffaloes and camels also fill this role.

Conservation. In the low rainfall areas of the DC (<600 mm per year), many of the plants consumed by grazing animals produce hard coated seeds. Left on the ground,two or more years would elapse before the seed would germinate. Passage of the seeds through animals scarifies the coat resulting in germination during the following rainy season. Without this cycle the ecological system would deteriorate more rapidly than is already occurring in arid and semi-arid regions, such as the Sahel region of Africa.

Small farm operators harvest their grains as soon as they can because they want animals to move on to the land to drop seeds with the hope there will be some plant growth arising. Even partial plant growth tends to reduce soil erosion during the fallow period. Ownership of animals frequently encourages or gives farmers purpose to plant a second or "winter crop", particularly legumes. In Egypt, about 60 percent of the crop land is planted to the legume Berseem clover during the winter months. This serves as the main source of animal feed from December to April and replenishes soil fertility.

<u>Pest Control</u>. Where crop cultivation is dependent on hand labor or tillage is by the use of primitive wooden plows, farmers want the land as free as possible of crop residues and weeds or grasses which have grown since the previous harvest. When there is mixing of pastoral herding and crop farming, animals are extensively employed to clear land for cropping. In parts of the Dominican Republic and Mexico, up to 30 percent of all feed for cattle and sheep is classed as land clearing.

Control of soil microorganisms is important in the performance of the newer varieties of wheat and rice in tropical areas. Rotation of crops with grasses or forage legumes, which can be used for animal feed, appears the most suitable means for "breaking the cycle" and provide a desirable environment for crops. In the paddy rice regions of southeast Asia, buffaloes are frequently employed to control grasses and weeds in the waterways. Buffaloes may also be driven through irrigation canals to crush snails that destroy crops and transmit disease to man (11).

<u>Cultural</u>. In addition to numerous roles which can be quantified in economic terms, animals which live with people become one thread in complex and highly involved social and cultural patterns. Animals are a source of identity and prestige or status for the family and a means of forming social ties through gifts and exchange with others. Barnett (19) well describes the cultural ties.

> "Pairs of yoked animals at the outset of the growing season pull plows and harrows, hours of work vis-a-vis acreage prepared may be measured and terms for rental of animal with or without driver, investigated. But it is only by living in the community throughout the full agricultural calendar and beyond that, that one begins to be sensitized to the more integrated role of livestock. They are linked with the life cycle of human beings. The small tractor might indeed replace the function of the buffalo in the field, they cannot do the same in the cultural context".

The role of animals in food supplies, or even traction, may be secondary to their part in recreation, religion and social custom. Rodeos and animal exhibitions, such as those seen at local or state fairs in the U.S., serve a role in recreation. Coleadas de Toros (throwing the bull by the tail) is a popular sport in Latin America. Bull fighting is well known in both Latin America and southeast Asia (11). Priangan sheep are bred primarily in Java for "ram fighting"

(20). Cock fighting is a village recreation throughout most of the DC. Equal to horse racing in history are races with buffaloes and cattle in certain of the DC.

More than one million of those of Islamic faith made the pilgrimage to Mecca (Saudia Arabia) during 1978. Most of these bought sheep, goats or poultry to provide the "great feast". In a number of societies, houses, roadways or bridges are inaugurated with a sacrificial ceremony utilizing one or more animals. The Montanard tribes of southeast Asia require sacrifice of a buffalo prior to planting their rice and repeat the ceremony post-harvest (1).

For almost the whole of Central Africa, animals are man's legacy to his sons. They are frequently used to symbolize formal contracts, such as marriage. Hundreds of examples of the roles of animals in cultural practices of various societies could be cited but this would detract from the central thrust; namely, that almost universally, societies give high priority to culturally defined needs in the utilization of animals. Frequently we overlook the need to take into account a delicate balance of economic productivity and cultural preference. The need to allow for these balances is especially important in the DC.

Inedible Products. Inedible products from animals are largely the creation of organized slaughtering among affluent societies. In the U.S., the largest product is inedible fats which are widely used for industrial purposes. In the DC, horns and hooves are the most valuable by-products. The core (pith) of horns and hooves contains a high percentage of gelatine which is used for glue in the making of furniture and handicrafts. The gelatine may also be used as food. Horns also frequently serve as the basis of cottage industries producing buttons, combs, handles and religious symbols. Preparation of sets of horns for use as religious symbols is a multimillion dollar industry in Malaysia, Indonesia, Philippines, Burma and Thailand. Trumpets prepared from the horns of rams are blown on Yom Kippur and have a sacred role in the Jewish religion.

When animals are slaughtered in the rural areas of the DC, portions of the carcass fat are removed for use as fuel for cooking or for the making of candles.

Conclusions

As we view the world situation, we find only a small percentage of domestic animals, about 1.5 percent, "competing" with humans for resources. The vast majority, quite to the contrary, exist symbiotically and provide man's only means of deriving life-sustaining products. In addition to converting inedible products into high quality protein, animals provide countless non-food uses which are essential contributions to the existence of humans. At this point in time we are unable to place a monetary value on contributions by animals through non-food services but these no doubt equal or exceed the value for meat, milk and eggs. This is especially the case in the DC. The recognition that fossil fuels are limiting, and that ecological considerations are important, is increasing the value of non-food services.

The U.S. model for livestock production focuses on deriving income from the sale of products which have value in the markets. This system can be characterized as the "meat/milk syndrome" since emphasis is primarily on production of these products in the most efficient manner with little or no regard for the output of other goods. Because of our major focus on meat or milk, we look upon the thin, boney looking cattle arriving at slaughter markets in the DC as having undergone gross mismanagement. However, when we recognize that the major strategy is to utilize animals for human subsistence, the emaciated animals entering the market can be viewed as having rendered valuable services rather well. The challenge for the DC is not one of shifting livestock to a meat/milk economy but rather working within traditional systems to increase the overall efficiency of the total services rendered by animals through such programs as health services. Even at this time, it is obvious we must use a different scale to access the role of animals in the DC.

References

1. R.E. McDowell, Ruminant products: More than meat and milk, Winrock Intern. Livestock Res. and Training Center, Morrilton, Ark. (1977).

2. R.E. McDowell, Are we prepared to help small farmers in developing countries? J. Anim. Sci., 46, 1184 (1978).

3. W.A. Verbeek, Proc. Second World Conf. Anim. Prod., Bruce Pub. Co., St. Paul, Minn., pp 61-77 (1969)

4. G. Beck, In: The role of animals in the world food situation, Rockefeller Foundation Rpt., New York, pp 24-27 (1975)

5. W. Allan, The African husbandman, Oliver & Boyd, Edinburgh (1965).

6. C. Oxby, Pastoral nomads and development: A selected annotated bibliography with special reference to the Sahel, Intern. African Inst., London (1975).

7. S. Sanford, Pastoral Network Paper 1C, Mimeo Overseas Dev. Inst., London (1975).

8. P.R. Baker, Nomadism in Africa, In: Animal husbandry in the tropics, 3rd Ed., G. Williamson and W.J.A. Payne Editors, Longman, London, p 720 (1978).

9. R.E. McDowell, Cornell Intern. Agri. Mimeo No. 57, p 1-3, Cornell University, Ithaca, N.Y. (1977).

10. R.E. McDowell, Role of animals in support of man, In: A series of papers on world food issues, Center for Analysis of World Food Issues, Program in Intern. Agri., Cornell Univ., Ithaca, N.Y. (1979).

11. W.R. Cockrill, The husbandry and health of the domestic buffalo, FAO, Rome, pp 304-08, 313-21, 324-26, 574-76, 639-40 (1974).

12. S.K. Barat, World Animal Rev. 14, 20 (1975).

13. D.H.L. Rollinson and A.J. Nell, The present and future situation of working cattle and buffalo in Indonesia, UNDP/FAO Proj., INS/72/009, FAO, Rome (1973).

14. A.U. Khan and J.B. Duff, Development of agricultural mechanization techniques at the International Rice Res. Inst., Paper 72-02, Seminar on the priorities for research on involving and adopting technologies for Asian development, Princeton Univ., Princeton, N.J. (June 5-7, 1972).

15. A. Makhijani and A. Poole, Energy and agriculture in the third world, Ballinger Pub. Co., Cambridge, MA. pp 15-60, 68-71, 79-86, 99-105, 114-20, 152-55 (1975).

16. U. Lele, Dev. Digest 14, 56 (1975).

17. N.S. Ramaswamy, The planning, development and management of animal energy resources in India, Occasional Paper No. 10, Indian Inst. of Mgt., Bangalore (1978).

18. V. Buvanendran, World Animal Rev., 27, 13 (1978).

19. M. Barnett, Livestock, rice and culture, Rock. Found. Rpt of Working Papers on Integrated crop and animal production to optimize resource utilization on small farms in developing countries, Bellagio, Italy (October 18-23, 1978).

20. I.L. Mason, World Animal Rev. 27, 13 (1978).

Vernon R. Young

7. Animal Foods, Past, Present and Future: A Nutritionist's View

Introduction

The raising of farm livestock and the production of animal food products are integral aspects of the agricultural industry in the U.S. and many other nations. Because the ultimate purpose of agriculture is to provide food to meet man's nutrient needs, as well as his desires, consideration must be given to the nutritional aspects of animal foods in any comprehensive and rational plan concerned with maximizing human food production. Furthermore, increased recognition is now being given to the nutritional health of the population as a key factor in national food, agriculture and health policies[1-3]. Although a comprehensive assessment of the role of animal agriculture and development of policies relating to the production and consumption of animal feeds and foods require integrated consideration of many geophysical, political, economic, environmental, social and other factors, it is no longer acceptable to debate and evaluate the future course of agricultural practices without involving human nutritional concerns.

In this brief paper, attention will be focused on the U.S. population for a number of reasons: First, because of animal agriculture and associated industries in the U.S.; and also because we often read or hear that we should reduce our intakes of beef, or full-fat milk, or animal fats, and of many highly desired foods. Second, many of the releveant research data on this topic have been generated in this country. Third, because we, as a nation, spend approximately 50% of our food dollar on products of animal origin (Table 1)[4]. Thus, changes in costs of animal foods and of meats, in particular, have significant effects on overall costs of food to the U.S. consumer. In the latter context, it is worth considering the impact of this current distribution of food-related expenditures on human nutritional status and

Table 1. Relative Importance of Major Food Groups to Per Capita Food Consumption Index.*

Group	1957-59	1967	1976
		%	
Meat	24.3	25.8	26.2
Fish	1.9	1.8	2.0
Poultry	4.2	5.5	6.0
Eggs	4.4	3.8	3.1
Dairy Products	19.0	16.9	16.1
Total Animal	54.6	54.5	53.8
Crop Products	65.4	45.5	46.2

*From Reference 4.

Table 2. Food Consumption Per Capita Per Year by Major Food Groups.*

Group	1909-13	1957-59	1977
Meat, Poultry, Fish			
Meat	141	144	160
Poultry	18	34	54
Fish	13	13	17
Eggs	37	47	34
Dairy Products (qts.)	177	239	225
Fats and Oils			
Butter	18	8	4
Total	41	49	59
Flour and Cereal Products	291	148	143
Sugars and Sweeteners	89	106	131

*Quantities in pounds except for dairy products. From: (6) Page and Friend and U.S.D.A. (4).

health and on what the future might hold. This evaluation will begin with a discussion of the importance of animal products in the U.S. diet. A description of trends in consumption of and contributions made by foods of animal origin to intakes of essential nutrients will be included. This will be followed by a short summary of factors that determine amounts of foods consumed. Then, the question of diet and human health, with particular reference to animal foods will be addressed. Finally, a brief account of recent developments in the area of nutrition, health and food policy will be presented. Important implications with respect to the distribution and type of animal foods that may characterize the U.S. diet during the coming decades will be identified.

Food Intake Patterns

Weir[5] discussed the several sources of data that provide information about the amount of animal products in the U.S. diet. These include national per capita estimates, household food-consumption surveys, the Ten-State Nutrition survey, the HANES survey, and surveys of selected sectors of the population and of individuals. For the present purpose, the national per capita estimates of the number of pounds of various food commodities available for use by the civilian population are useful to examine. These data are developed annually by the U.S. Department of Agriculture (USDA) and the former represent "disappearance data", because they do not take into account losses or waste after food leaves the retail outlet. They do not provide information on the distribution of food among sectors of the population or within families. However, these data are valuable because they have been collected annually since the early part of this century and, thus, they serve as an index of approximate per capita consumption, of trends in animal product intake and as an index of the nutritional adequacy of the national diet.

As shown in Table 2, intake of meats in 1909-13 was almost as high as it was in 1977. There has been a marked rise in poultry consumption but, in contrast, a distinct decline in consumption of flour and cereal products has occurred during this same period. Sugars and sweeteners, have risen markedly in use. An important point to be drawn from these data is that there have been several simultaneous trends in the types of foods consumed with decreased intakes of some foods compensating increased intakes of others.

The apparent constancy of total meat available for consumption, shown in Table 2, does not reveal changes in proportion of various meats available for consumption in the

Table 3. Consumption (Lb/Capita) of Meats For Period 1960-1976 *

Year	Beef	Veal	Pork	Lamb & Mutton	Offal	Canned Meat†
1960	64.3	5.2	60.3	4.3	10.1	10.8
1963	69.9	4.1	61.1	4.4	10.3	11.8
1966	77.1	3.8	54.3	3.6	10.3	12.4
1969	82.0	2.7	60.6	3.0	10.7	14.3
1972	85.9	1.8	62.9	2.9	10.7	14.0
1976	95.6	3.3	54.5	1.7	10.7	13.1

*Retail cut equivalent; †Net canned weight.
Source: U.S.D.A. (4).

Table 4. Nutrients Available for Consumption, Per Capita, for 1977 in Relation to the Previous Two Decades *

Nutrient	Unit	1977	1977 as % of:		
			1957-59	1967	1976
Energy	Kcal	3380	108	105	100
Protein	g	103	108	104	100
Fat	g	159	111	106	100
Carbohydrate	g	391	104	105	100
Calcium	g	0.94	96	100	99
Iron	mg	18.60	115	107	100
Vitamin A	I.U.	8200	101	103	100
Ascorbic Acid	mg	116	111	111	98

*From: National Food Review, January, 1978, U.S.D.A., S.E.A., Hyattsville, Maryland (7).

U.S. These are summarized in Table 3. A consistent rise has occurred over the past two decades in beef consumption; approximating a 50% increase[7]. In contrast, intakes of veal, mutton and pork have declined. Consumers show a preference for pork over veal and mutton, but the first choice is red meat. A detailed consideration of base for these apparent changes in preference and for the high degree of acceptability of beef is beyond the scope of this review. However, Pearson [8,9] identified and evaluated some of the factors including excessive fattness and off-flavors in lamb, and preferences for the colors of lean and fat in beef. A major factor in the increase in beef consumption may be the demand for ground meat products by the fast-food industry.

From a nutritional standpoint, these changed patterns in food consumption have caused a redistribution among sources of nutrients, particularly for the energy-yielding nutrients, carbohydrates and fats, and for proteins. These are depicted in Figure 1. Thus, shifts in the food sources and types of energy-yielding nutrients have occurred. During the period 1909-1913, grain products, as a food group, were the largest contributor to energy intake, accounting for more than twice that of other foods including meats and dairy products. Today, the contribution of grain products has declined to about one half of total energy intake and approximates the contribution made by meats, fats and oils. Furthermore, as revealed in Figure 1, there has been a predictable shift in the source of dietary protein. Animal products now provide more than two thirds of dietary protein as compared with equal contributions of animal and vegetable proteins in 1909-13. Similar trends in food habits have been described for Canada[10] and several European countries[11]. Thus, it can be anticipated that changes in food choices have had an impact on the relative contributions made by animal foods to total intakes of the about forty-five essential nutrients.

According to the USDA figures (Table 4), the "average" diet in 1977 provided energy, protein and other nutrients at levels slightly in excess of those for 1957. With the exception of specific nutrients for some groups of individuals, the evidence suggests that nutrients available for consumption usually exceeds current recommended daily allowances[12]. For a number of these nutrients, in addition to energy and protein (Figure 1), meat, eggs, and dairy products make an important or major contribution to their total daily intake. The data (Table 5) reveal that these animal foods account for 50% or more of the total intake of many B vitamins, 38% of iron, 81% of calcium and more than 90% of vitamin B_{12}. The latter is not synthesized by the plants we

Table 5. Contribution of Major Animal Food Groups in 1977 to Nutrient Supplies Available for Civilian Consumption.*

Food Group	Energy	Protein	Fat	Iron	Calcium	Vitamins A	Vitamins B_{12}	Vitamins C
	%							
Meat, Poultry, Fish	20	43	34	31	4	22.4	70.5	1.1
Eggs	1.8	4.8	2.7	4.7	2.2	5.5	7.9	0
Dairy Products (-butter)	11	22	12.5	2.5	74.6	13.0	20.1	3.9
Fats and Oils (+butter)	18	0.2	43	0	0.4	8.3	0	0
TOTAL	51	70	92	38.2	81.2	49.2	98.5	5.0

*Abstracted from: Reference 7.

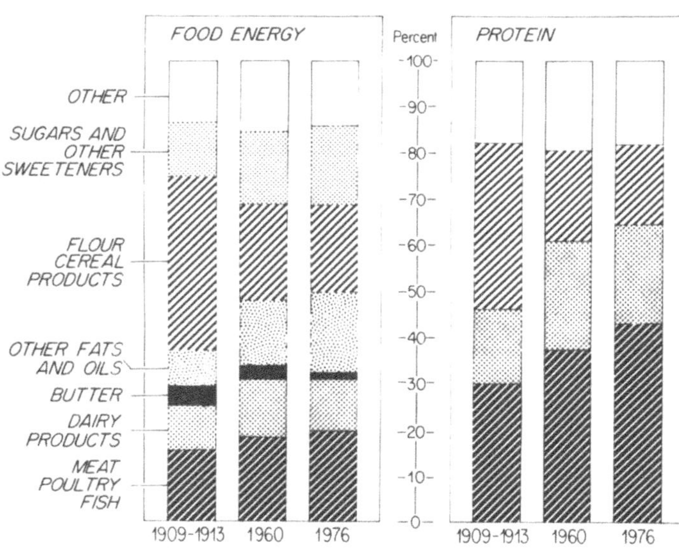

Figure 1. Percentage of energy and protein contributed by major food groups.

consume as foods. For some vitamins, such as vitamin C, the contribution of animal foods is relatively small. Nevertheless, the high availability of iron[13,14] and other minerals[15], the good quality of the protein for meeting human essential amino acid requirements[16], and the relatively high concentration of many vitamins in animal foods indicate that they function as a significant source of nutrients for humans[18].

It should be emphasized that there is no absolute need for animal protein or food of animal origin, at least for achievement of intakes comparable with those in the current U.S. diet. A series of diets consumed by various populations and/or ethical groups, which contain minimal or essentially zero levels of nutrients for growth and maintenance of normal organ and tissue function are summarized in Table 6. It should be stressed that diets based entirely on a narrow range of plant foods may be marginal or deficient in some nutrients such as B_{12} and calcium. Supplements or a small amount of animal foods can correct this inadequacy[19].

Factors Determining Food Choice

Some of the factors that may be responsible for the trends discussed above and present intakes of animal foods by the U.S. population should be identified at this point. These are listed in Table 7. The most important factor is bioavailability. It is obvious that a food is not eaten when it is not available. Advances in food processing and technology and in the distribution and marketing of foods have markedly improved accessability of uncontaminated, acceptable foods throughout the year. Of course, it is necessary to be able to pay for foods. Thus, income is a determinant of amounts and types of foods consumed. Income has been clearly shown to play a significant role in food selection[20]. Many studies have suggested that increases in real income, at least up to a threshold level[21], result in a shift from cereal based diets to diets in which animal-derived foods predominate. Social and cultural factors have been discussed extensively by others[22]. Finally, there are physiological determinants of man's food choice, Yudkin[22], suggested that palatability and, perhaps, specific satiety help direct us to diets that are adequate in quality and quantity. This is not to say that we are capable of achieving a balance between what we want and what is necessary for physiological needs and optimal health maintenance. The obesity, hypertension, heart disease and dental caries so prevelant in our society are examples of the imbalance among nutrients so many of us select[23].

A number of specific issues particularly relevant to

Table 6. Examples of Adequate Near-Non-Flesh and Non-Flesh Diets of Population Groups Reported in Literature.*

Investigator	Population Group	Characteristics of Diet
McCarrison	Hunza	wheat, barley, millet, maize, legumes, vegetables, milk, meat occasionally.
Richards	Bemba	finger millet, maize, sweet potatoes, legumes, plantains, vegetables, little animal food.
Steiner	Okinawans	rice, sweet potatoes, soybeans, some milk, vegetables, meat infrequently.
Adolph	N. Chinese	wheat, millet, barley, corn, soybeans, other legumes, vegetables.
Anderson et al	Otomi Indians, Mexico	corn (mainly as tortillas, up to 80% of total calories), legumes, vegetables.
Toor et al	Yemenite Jews	large quantities of "pita" (flat bread) and vegetables, sunflower seeds, legumes, nuts, little meat.
Walker	South African Bantu	50-90% whole-ground or lightly milled cereals, corn, sorghum and wheat, legumes, vegetables and greens; a little milk, eggs and meat.
FAO/WHO Report	Lebanese	56% of calories from cereals-wheat,barley, millet, rice; milk, cheese, legumes, fruits, vegetables, olive oil, and ghee; meat intake very low.

*From Reference 19.

assessments of future trends in the consumption of animal foods appear in this list of factors presented in Table 7. Desire by consumers to purchase products for either implied health needs or expected benefits is increasing. Scala[24] (Table 8) summarized trends in the purchase of low-tar cigarettes and consumption of skim milk and yogurt. The data imply that consumers are seeking alternatives perceived to reduce risk of disease or produce a more vigorous state of nutritional health. The increasing number of books concerned with diet and nutrition on the market is further evidence of this. Unfortunately, as well stated by Clydesdale[25], "too often these books provide prestige and wealth to the authors, but little more than nutrition fantasy for their readers."

Introduction of nutrition labelling of foods[26] is potentially a tool for increasing the visibility of nutrition to the consumer, as well as serving as a vehicle for nutrition education. A recent survey has shown that a significant number of shoppers are aware of nutrition labelling (Table 9). Some consumers claim to utilize information given on food labels to choose foods which may offer a nutritional advantage[27]. However, improvements must be made in nutrition labelling before sufficient numbers use the nutrition information to help in planning nutritionally sound diets.

Animal Foods and Health Concerns

In view of the earlier point that animal foods account for a significant proportion of total daily intakes of many essential nutrients, it becomes important to assess contributions made by animal foods to nutrients often consumed in considerable excess of needs; particularly, lipid (fat) and protein. This is pertinent because in the U.S., and other technically advanced nations, the development of diseases of major public health concern, shown in Table 10, have been connected with excess and imbalances of nutrient intakes.

Epidemiological, clinical and experimental animal studies lead to the conclusion that high intakes of saturated fat and cholesterol are risk factors in coronary heart disease[28-32]. Because animal products contribute about 50% of total available fat for consumption (Table 11) and a major fraction of daily cholesterol intake, is worth considering briefly this aspect of the use of animal foods.

Nutrient fat available for consumption has increased from 125g in 1909 to 157g per capita daily in 1977 (Figure 2)[33]. Even though this increase is largely due to increased consumption of vegetable fats and separated oils (Figure 3), animal fats continue to account for an important share of total fat

Table 7. Some Factors Modifying Food Patterns and Habits

1. Availability
2. Income
3. Cultural and Social
4. Physiological Determinants

Table 8. Trends in Consumption of Health-Oriented Products*

Year	Low-Tar cigarettes (billions)	Low-fat and skim milk (gal/capita)	Yogurt (oz/capita)
1960		13	0.26
1965	13	15	0.32
1970	27	43	0.86
1975	65	61	1.64

*From: Scala (24).

Table 9. Nutrition Labelling: Consumer Response.*

Age Group	% of food shoppers who:	
	have noticed	have used
— Awareness and Use —		
18 - 34	72	43
35 - 49	63	32
50+	45	25
TOTAL	58	33

Intended Use	% of food shoppers
Help get best nutrition buys	42
Help plan better home diet	28
Will not use too much	22

*From: Schrayer (27).

Table 10. Diseases in Which Diet is Considered to be of Causative Significance or Acts as a Modulator.

Obesity
Dental Caries
Hypertension
Diabetes
Coronary Heart Disease
Cancers

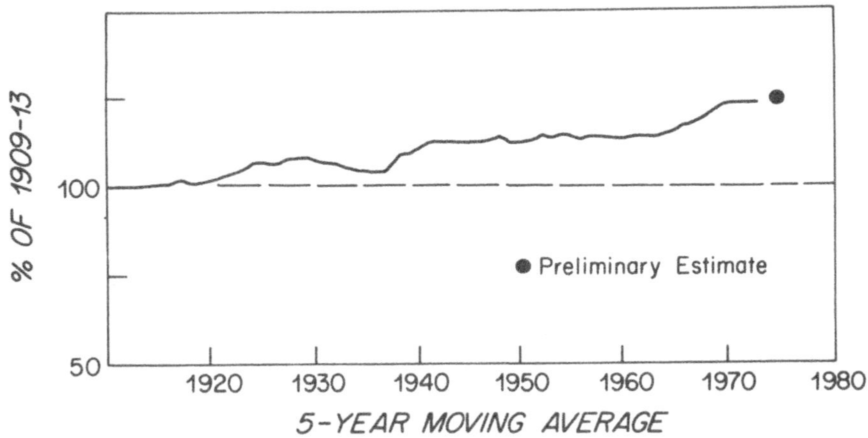

Figure 2. Per capita consumption of nutrient fat.

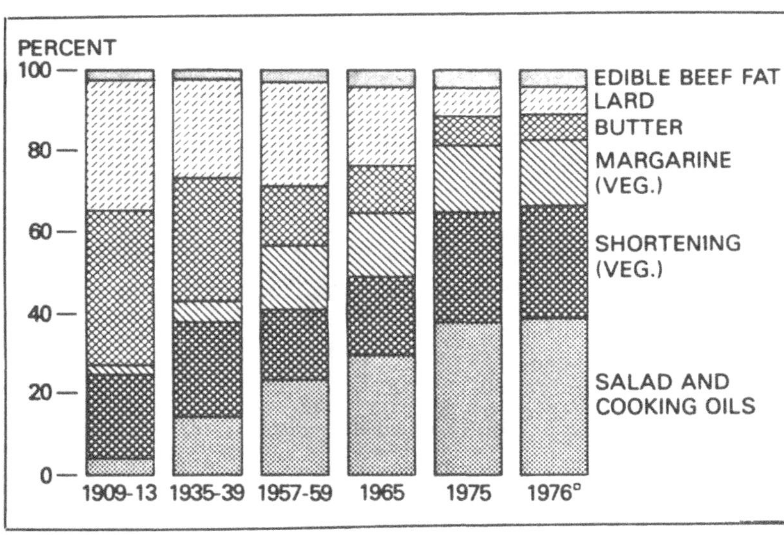

Figure 3. Nutrient fat from fats and oils.

intake. Furthermore, a major portion of saturated fatty acids available for consumption is contributed by animal products.

Epidemiological studies have established strong correlations between intake of animal products, saturated fat, and cholesterol, and incidence of coronary heart disease[28-32] (Table 12). Similarly, risk of breast cancer seems to be related to fat intake[33-35]; (Table 13), but the relative importance of animal and vegetable fats in breast cancer is still debatable [36-37]. Also, meat proteins have been postulated to increase risk of intestinal cancers, possibly by an action of meat extracts on intestinal microflora that can produce carcinogenic substances [38-39].

In relation to future food uses of animal products, it is important to determine whether these various correlations between dietary variables and disease incidence indicate etiology significant associations. Because of the complexities both of human cancers and of diet patterns, each with its own set of internal correlations among nutrients and each with a set of associations with non-dietary components of life-style[40], critical interpretation of epidemiological data is a formidable task. Stamler[28] proposed a number of criteria (Table 14) to help assess whether epidemiological associations are significant in etiology. Based on these criteria, it is reasonable to conclude from the available data that high intakes of saturated fat and, less importantly, of cholesterol,increase the risk of heart disease while lower intakes reduce this risk.

The data base available for analysis of relationship between animal products and human cancers is less extensive than that for dietary factors and coronary heart disease. Thus, it is premature to conclude that animal proteins, per se, alter risks of various forms of cancer in humans. There is, however, a related issue worthy of comment. This is the use of nitrites in cured meats and for formation of nitrosamines which are putative carcinogens[41]. This is a bit beyond the general scope of the present review, but of significant current interest. Tannenbaum et. al. [42] have shown that there is a considerable endogenous synthesis of nitrite within the human gastrointestinal tract. There is a greater output of nitrate in the urine than is present in diets devoid of nitrite and low in nitrate (Table 15). Based on these results, it was estimated that endogenous nitrite synthesis far exceeds nitrite likely to be consumed as an intentional additive to variety meats. Thus, the

Table 11. Percentage Contribution to Fat Consumption by Various Food Groups.*

Group	1909-13	1960	1976
Meat	34.0	31	30
Poultry	1.8	1.7	2.2
Dairy Products	14.9	15.5	12.5
Butter	14.2	5.3	2.8
Fats and Oils (-butter)	36.9	39.6	44.0

* From U.S.D.A. (7).

Table 12. Food Groups and Nutrients Per Capita, 1954-65 and Coronary Heart Disease Mortality Rate, 20-Country Study.†

Food or Nutrient	Simple r CHD Mortality Rate	
	Men	Women
Meat, poultry, dairy eggs	0.714***	0.528**
Dairy products and eggs	0.698***	0.465*
Sugar	0.748***	0.671***
Calories/day	0.635**	0.475*
Cholesterol (mg/day)	0.655***	0.561**
Saturated fat (g/day)	0.676***	0.476*
Protein (g/day)	0.674***	0.556**

† Combined from Stamler (28); *,**,*** $p< 0.05, 0.01, 0.001$, respectively.

Table 13. Correlations Between Endocrine Cancers and Food Consumption Patterns in 37 Countries.*

	Pearson's r
Breast Cancer	
and total fat	∿ 0.80
combined fat	∿ 0.80
animal protein	∿ 0.75
animal fat	∿ 0.75
eggs	∿ 0.75
total protein	∿ 0.60
sugar	∿ 0.50

* From Berg (34), based on Drasar and Irving (35).

Table 14. Guidelines for Assessing Etiological Significance of Epidemiological Associations.*

1. Strength of the association
2. Graded nature of the association
3. Temporal sequence
4. Consistency of finding
5. Independence of each of the associations
6. Predictive capacity
7. Coherence of findings (with animal, clinical and pathological mechanisms).

* From Stamler (28).

Table 15. Mean Daily Intake and Urinary Excretion of Nitrate Nitrogen in Six Adult Subjects.*

Diet	mg Nitrate N/person/day	
	Ingested	Excreted
"Protein-free"	0.94	17.70
Egg Protein (0.8 g/kg/day)	1.76	16.51

* Summarized from: Tannenbaum et al. (42)

Table 16. Guidelines for Success in Bringing About Change in Food Habits.*

1. Work within current changes in consumer food habits, wherever possible.
2. Identify consumer attitudes and needs and their ramification on future choice.
3. Incorporate changes within existing food patterns.
4. Know what effect of change in food habit is likely to be on overall behavior patterns, food habits, cooking procedures, and extent of social and economic change.
5. Work within current environment.
6. Government regulations of great importance.
7. Approach must be viewed as a realistic economic operation.

* Based on McKenzie (43).

issue of nitrite and cancer risk is more complex than some assume.

Returning, however, to saturated fat and cholesterol; if these diet components increase the risk of coronary heart disease and if a reduction in their intake is in the health interest of consumers, it is important to consider the possible impact of this observation on future consumption of animal products.

McKenzie[43] reported that desirable changes in food habits are best achieved by following the guidelines summarized in Table 16, and a number of guidelines are particularly relevant to this discussion. For example, chemical modification of ruminant diets has been tried as a basis for altering the distribution of fatty acids in foods from ruminants[44,45]. Ruminants fed diet containing polyunsaturated fats coated with formalin treated proteins to protect from microbial fermentation in the rumen, produced beef and milk fats significantly higher in linoleic acid. Would this provide any important nutritional advantages for man? Based on available data, Heywood[45] concluded that these products would have only a small influence on serum cholesterol and, therefore, a minimal advantage in preventive health. It seems that a more effective health benefit would result from reducing total saturated fat intake by choosing leaner cuts of meat, and trimming visible fat. Indeed, the demand for reduced fatness is not new[8]. This goal can be aided by breeding programs, changes in management, and perhaps less significantly by alterations in standards of grading meat and other animal products[46], Finally, it must be emphasized that fat and cholesterol are not the only risk factors of importance in heart disease. Total food intake, physical activity, life styles and cigarette smoking are all factors [47,48]. Nevertheless, present evidence identifies saturated fat and cholesterol as major dietary risks factors leading to the conclusion that it should be prudent to moderate intakes of these to reduce risk from developing heart disease.

Vegetable Proteins

Different types of food proteins appear to differ in artherogenicity [49,50]. Studies [50] in rabbits at equivalent fat and cholesterol intakes have shown that proteins of plant origin support lower levels of blood cholesterol than do various sources of animal protein. Although the public health significance of this finding remains to be determined, it does bring into focus, for this discussion, consideration of increasing utilization of vegetable proteins in combination with animal products and with meat in particular.

Table 17. Projected Use of Vegetable Protein as Meat Extenders and Impact on Animal Herd.*

Meat Replacement	
Low	10%
Medium	16%
High	21%
Animals replaced (%)	
Cattle, hogs, sheep	6.5
Chicken	0.2
Turkey	2.4

* Source: Juls (53).

Table 18. Summary of N Balance Data and Protein Quality Indices in Soy-Beef Replacement Study in Young Adult Men.*

% Soy	100	75	50	25	0
% Beef	0	25	50	75	100
N Balance+	-2.3	-3.2	-0.9	-1.1	-1.7
BV++	53	52	55	53	53
Digestibility (%)	97	99	98	98	98

*From reference 55; + mg N/kg/day; ++ biological value.

None of the diets showed significant ($p > 0.05$) differences.

Vegetable proteins have been used for various purposes in association with animal foods for some time. Recently, however, there has been a trend towards an increased use for various animal based foods[51,52]. It has been estimated[53] that their use for this purpose will increase further and that this could have a significant impact on the size of livestock herds (Table 17). These trends offer potential nutritional and economic advantages to the consumer and if this is so, they should be encouraged. Furthermore, we[54,55] have shown in human metabolic studies, that soy protein can extend beef without a detectable change in overall protein quality (Table 18). Studies by other investigators have shown this also applies in children and that well-processed soy protein can meet adequately, the essential amino acid needs of subjects of all ages [55].

The development of various meat extendors and replacers may be regarded as a desireable change for the consumers since these products may offer, not only, health advantages but also result in an increased variety of potentially less expensive and acceptable foods. Athough there is still a need for more nutritional and food science research before the full impact of vegetable protein sources in the U.S. diet is achieved, it is apparent that these products are likely to influence overall intake of animal foods by the U.S. consumer in the coming years. However, this need not be viewed as a major threat to overall animal food production, particularly as the U.S. population continues to increase and constraints on food resource allocation became more critical for the U.S. and the world.

Food Nutrition and Policy

In any overview of animal products, from a nutritionist's viewpoint, it is necessary to consider food and nutrition policies. The long-term nutritional health of some greater than 200 million Americans can only be assured through responsible government action[56], Furthermore, it is evident that policy makers are now increasingly aware of linkages between dietary practices, nutrition and health. The important need to develop a sound national nutrition and food policy for this country has been discussed[57,58]. The role of diet in preventive medicine must be fully recognized. It cannot be said that the dietary trends, discussed earlier or that present diets are necessarily physiologically appropriate or optimal for long-term health maintenance.

During the past few years, the Senate Select Committee on Nutrition and Human Needs, now the Subcommittee on Nutrition, Senate Agriculture, Nutrition and Forestry

Table 19. U.S. Dietary Goals.*

1. To avoid overweight, consume only as much energy (calories) as is expended; if overweight, decrease energy intake and increase energy expenditure.

2. Increase the consumption of complex carbohydrates and "naturally occurring" sugars from about 28% of energy intake to about 48% of energy intake.

3. Reduce the consumption of refined and processed sugars by 45% to account for about 10% of total energy intake.

4. Reduce overall fat consumption from approximately 40% to about 30% of energy intake.

5. Reduce saturated fat consumption to account for about 10% of total energy intake; and balance that with polyunsaturated and monounsaturated fats, which should account for about 10% of energy intake each.

6. Reduce cholesterol consumption to about 300 mg/day.

7. Limit the intake of sodium by reducing the intake of salt to about 5 g/day.

* Dietary goals for the United States (December, 1977).

Committee, has been active in assessing relationships between nutrition and health issues. In December, 1977[59], a revised set of seven dietary goals for the U.S. were presented. These goals are depicted in Figure 4 and stated more fully in Table 19. Included are recommendations to avoid overweight, to increase consumption of complex carbohydrates, to reduce intake of refined sugars, to reduce the consumption of total fat, saturated fat and cholesterol, and to limit salt intake. At this time, the goals are intended to serve as guidelines for individuals. They do not act as a specific basis for legislative policy.

To achieve these dietary goals, a series of changes in food selection have been proposed (Table 20). As can be seen, these suggestions are relevant to the future use of animal foods, particularly as they concern the intake of animal fat and lean meats. Furthermore, proposals have been made with regard to choices of dairy products. If these changes in food intake patterns become a matter of policy, they will have an obvious impact of future amounts of animal foods in the U.S. diet. The impact will depend upon the specific food choices and changes in food habits made in attempting to meet these goals[60-61].

There has been considerable debate regarding the specificity and need for these dietary goals [e.g.62,63]. It is not likely that they are necessary for all people and there is room for refinement. Also, there is a genuine need to further evaluate their adequacy. Determination of whether or not suggestions of the kind outlined in the U.S. dietary goals will significantly reduce the incidence of degenerative diseases, including cancers, diabetes, and heart disease, will require implimentation of a sound nutritional, nationally-oriented, surveillance program.

Lee[64] has pointed out that the national debate on nutrition and food issues has been initiated and that the U.S. Dietary Goals represent a significant step toward the development of a sound national nutrition policy for the few years ahead. Similarly, Moragne[65] stated, "work in nutrition and public policy has just begun." Furthermore, I agree with Jolly[66], who states, "If the health and nutritional status of the total U.S. population is to be protected, we must set nutritional achievement as an explicit objective of the food supply system. The days of major nutritional deficiency diseases are behind us and we must now consider balance and levels of nutrient and food intake that are consistent with health maintenance within the context of our increasingly complex total environment".

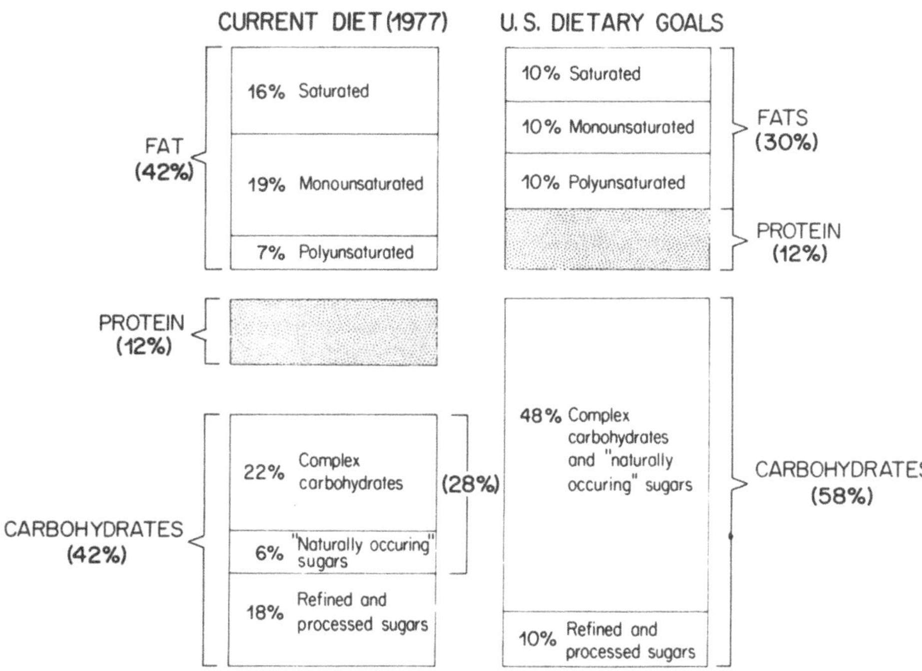

Figure 4. Percentage of total calorie intake for major dietary energy sources.

Table 20. Changes in Food Selection That Are Suggested in Meeting the U.S. Dietary Goals.*

1. Increase consumption of fruit and vegetables and whole grains.

2. Decrease consumption of refined and other processed sugars and foods high in such sugars.

3. Decrease consumption of foods high in total fat, and partially replace saturated fats, whether obtained from animal or vegetable sources, with poly-unsaturated fats.

4. Decrease consumption of animal fat, and choose meats, poultry and fish which will reduce saturated fat intake.

5. Except for young children, substitute low-fat and non-fat milk for whole milk, and low-fat dairy products for high fat dairy products.

6. Decrease consumption of butterfat, eggs, and other high cholesterol sources. Some consideration should be given to easing the cholesterol goal for pre-menopausal women, young children and the elderly in order to obtain the nutritional benefits of eggs in the diet.

7. Decrease consumption of salt and foods high in salt content.

---

* From: Dietary Goals for the United States (December, 1977).

Table 21. Some Pros and Cons of Animal Products in Relation to Human Nutrition.*

Pro	Con
1. Excellent acceptance	1. High cost
2. High protein quality	2. High perishibility
3. Excellent source of nutrients (vitamins, minerals)	3. Concerns in relation to health
4. Flexibility in meal planning	

* From reference 67.

Summary and Conclusion

In this overview, an attempt was made to highlight important nutrition and health issues that our society faces and to relate these to present and future uses of animal foods in the U.S. diet.

Animal products have a number of positive attributes for human nutrition (Table 21). However, actual or potential disadvantages arising from an increased proportion of animal foods in the diet must be recognized. At this time, not all of the disadvantages are easily quantified nor are they accepted by all health professionals. Nevertheless, they must be carefully and responsibly faced by the agricultural, animal and food industries because they each can influence patterns of consumption of animal products in the near and long-term future.

It is not possible to predict reliably the future role of animal foods in human nutrition. At present, it appears that with some relatively modest changes in the types of animal products made available to the consumer, advances in food technology and moderation in amounts of these and other foods consumed, the quantitatively important role of animal products in the U.S. diet should remain for at least the next two decades. Whatever the long-term outcome, it is now clear that concerns for the nutrition-health axis must be taken into account as factors in determining the production and consumption of various animal products during the years ahead.

References

1. Hegsted, D.M., 1973. J. Am. Dietet. Assoc. 62:394.

2. Foreman, C.T., 1978. Food Policy 3:216.

3. Select Committee on Nutrition and Human Needs, 1977. Dietary Goals for the United States. U.S. Government Printing Office, Washington, D.C.

4. U.S. Department of Agriculture (U.S.D.A.), 1978. Economics, Statistics, and Cooperative Service Supplement for 1976 to Agricultural Economic Report No. 138, pp. 76. U.S.D.A., Washington, D.C.

5. Weir, C.E., 1976. In: Fat Content and Composition of Animal Products. p. 5-23. National Academy of Sciences, Washington, D.C.

6. Page, L. and Friend, B., 1978. Bioscience 28:192.

7. U.S. Department of Agriculture, 1978. National Food Review Science and Educ. Admin. Consumer and Food Econ. Institute Maryland.

8. Pearson, A.M., 1976. In: Fat Content and Composition of Animal Products. p. 45-79. National Academy of Sciences, Washington, D.C.

9. Pearson, A.M., 1976. J. Am. Diet. Assoc. 69:522.

10. Campbell, J.A., 1978. In: Diet of Man: Needs and Wants (ed. J. Yudkin) p. 47-63. Applied Science Publishers Ltd., London.

11. Elton, G.A.H., 1978. In: Diet of Man: Needs and Wants (ed. J. Yudkin) p. 25-40. Applied Science Publishers Ltd., London.

12. Chopra, J.G., Forbes, A.L., and Habicht, J.P., 1978. J. Diet. Assoc. 72:253.

13. Layrisse, M., Cook, J.D., Martinez-Torres, C., Roche, M., Kuhn, M., Kuhn, I.N., Walker, R.B., and Finch, C.A., 1969. Blood 33:430.

14. Cook, J.D., 1977. Fed. Proc. 36:2028.

15. Evans, G.W., and Johnson, P.E., 1977. Am. J. Clin. Nutr. 30:873.

16. Young, V.R., Fjardo, L., Murray, E., Rand, W.M. and Scrimshaw, N.S., 1975. J. Nutr. 105:534.

17. FAO/WHO, 1973. Energy and Protein Requirements, World Health Organization Technical Rept. Ser. #522, Geneva, Switzerland.

18. Munro, H.N., 197. In: Fat Content and Composition of Animal Products, p. 24-44. National Academy of Sciences, Washington, D.C.

19. Register, U.D., and Sonnenberg, L.M., 1973. J. Am. Diet. Assoc. 62:253; Sanders, T.A.B., 1978. Plant Foods for Man. 2:181.

20. Altschul, A.A., 1974. In: New Protein Foods Vol. 1A (ed. A. Altschul), p. 1. Academic Press, New York.

21. Miller, S.A., 1978. In: Diet of Man: Needs and Wants (ed. J. Yudkin), p. 187-204. Applied Sciences Publishers Ltd., London.

22. Yudkin, J., 1978. In: Diet of Man: Needs and Wants (ed. J. Yudkin), p. 243-256. Applied Sciences Publishers Ltd., London.

23. Winick, M., 1978. Food Tech. 32:42; Gori, G.B. and Richter, B.J., 1978. Science 200:1124.

24. Scala, J., 1978. Food Tech. 32:77.

25. Clydesdale, F.M., 1978. Food Tech. 32:127.

26. Rusoff, I.I., 1978. Food Tech 32:33; Forbes, A.L., 1978, Food Tech. 32:37.

27. Schrayer, D.W., 1978. Food Tech. 32:42.

28. Stamler, J., 1978. Circulation 58:3.

29. Stamler, J., 1979. In: Nutrition, Lipids, and Coronary Heart Disease (Eds. R. Levy, B. Rifkind, B. Dennis, and N. Ernst), p. 25-88. Raven Press, New York.

30. Glueck, C.J., and Connor, W.E., 1978. Am. J. Clin. Nutr. 31:727.

31. Kritchevsky, D., 1976. Am. J. Pathol. 84:615.

32. Heyden, S., 1975. In: The Role of Fats in Human Nutrition (ed. A.J. Vergroesen), p. 44-113. Academic Press, New York.

33. Enig, M.G., Munn, R.J., and Keeney, M., 1978. Fed. Proc. 37:2215.

34. Berg, J.W., 1975. Cancer Res. 35:3345.

35. Drasar, B.S., and Irving, D., 1973. Brit. J. Cancer. 27:167.

36. Caroll, K.K., 1975. Cancer Res. 35:3374.

37. Caroll, K.K., and Khor, H.T., Progr. Biochem. Pharmacol. 10:308.

38. Wynder, E.L., 1976. Fed. Proc. 35:1309.

39. Wynder, E.L., 1977. J. Am. Diet. Assoc. 71:385.

40. Knox, E.G., 1977. Brit. J. Prevent. Soc. Med. 31:71.

41. Lijinsky, W., and Epstein, S., 1968. Nature (Land) 225:21.

42. Tannenbaum, S.R., Felt, D., Young, V.R., Land, P.D. and Bruce, W.R., 1978. Science 200:1487.

43. McKenzie, J.C., 1977. Proc. Nutr. Soc. (Engl.) 36:317.

44. Oltjen, R.R., and Dinicus, D.A., 1975. J. Anim. Sci. 41:703.

45. Heywood, P.F., 1977. Am. J. Clin. Nutr. 30:1726.

46. Pierce, J.C., 1976. In: Fat Content and Composition of Animal Products, p. 183-188, National Academy of Sciences, Washington, D.C.

47. Shank, R.E., 1979. In: Nutrition, Lipids and Coronary Heart Disease (eds. R. Levy, B. Rifkind, B. Dennis and N. Ernst). p. 523. Raven Press, New York.

48. Levy, R.I., Rifkind, B.M., B. Dennis, and N. Ernst. Nutrition, Lipids and Coronary Heart Disease. Vol. 1, pp. 566. Raven Press, New York.

49. Kritchevsky, D., 1979. J. Am. Oil. Chem. Soc. 56:135.

50. Caroll, K.K., and Hamilton, R.M.G., 1975. J. Food Sci. 40:18.

51. Kinsella, J.E., 1979. J. Am. Oil Chem. Soc. 56:242.

52. Gallimore, W.W., 1979. J. Am. Oil. Chem. Soc. 56:181.

53. Jules, M., 1975. Danish Meat Products Laboratory, Manuscript No. 107., Royal Vet. and Agric. Unit, Denmark.

54. Scrimshaw, N.S., and Young, V.R., 1979. In: Soy Protein and Human Nutrition (ed. H.L. Wilcke, D.T. Hopkins, and D.H. Waggle) p. 121-143. Academic Press, New York.

55. Young, V.R., Scrimshaw, N.S., Torun, B., and Viteri, F., 1979. J. Am. Oil. Chem. Soc. 56:110.

56. See for example, Gori, G.B. and Richter, B.J., 1978. Science 200:1124. Priority of funding, resource allocation, training and legislation, clearly involve major government organizations.

57. Hegsted, D.M., 1978. Food Tech. 32:44.

58. Mayer, J., 1973. In: Nutrition Policies in the Seventies (ed. J. Mayer). W.H. Freeman Co., San Francisco.

59. Senate Select Committee on Nutrition and Human Needs. 1977. Dietary Goals for the United States (2nd Edition). Dec. 1977, U.S. Government Printing Office, Washington, D.C.

60. Peterkin, B., 1978. Food Tech. p. 34. February.

61. Peterkin, B., Shore, C.J., and Kerr, R.L., 1979. J. Am. Diet. Assoc. 74:423.

62. Hegsted, D.M., 1978. Am. J. Clin. Nutr. 31:1504.

63. Harper, A.E., 1978. Am. J. Clin. Nutr. 31:310.

64. Lee, P.R., 1978. J. Am. Diet. Assoc. 72:581.

65. Moragne, L., 1978. Food Tech. 32:54.

66. Jolly, D.A., 1976. J. Nutr. Educ. 8:56.

67. Binkerd, E.F., Forsythe, R.H., Hendrickson, R.L., Ives, R.J., Kolari, O.E., and Tracy, C., 1978. In: Protein Resources and Technology: Status and Research Needs (ed. M. Milner, N.S. Scrimshaw and D.I.C. Wang). p.389-42. AVI Publishing Co., Inc., Wesport, Conn.

For Product Safety Concerns and Information please contact our EU representative GPSR@taylorandfrancis.com Taylor & Francis Verlag GmbH, Kaufingerstraße 24, 80331 München, Germany